A Beginner's Guide to Mathematical Proof

A Beginner's Guide to Mathematical Proof prepares mathematics majors for the transition to abstract mathematics, as well as introducing a wider readership of quantitative science students, such as engineers, to the mathematical structures underlying more applied topics.

The text is designed to be easily utilized by both instructor and student, with an accessible, step-by-step approach requiring minimal mathematical prerequisites. The book builds towards more complex ideas as it progresses but never makes assumptions of the reader beyond the material already covered.

Features
- No mathematical prerequisites beyond high school mathematics
- Suitable for an Introduction to Proofs course for mathematics majors and other students of quantitative sciences, such as engineering
- Replete with exercises and examples.

Mark DeBonis received his PhD in Mathematics from the University of California, Irvine, USA. He began his career as a theoretical mathematician in the field of group theory and model theory, but in later years switched to applied mathematics, in particular to machine learning. He spent some time working for the US Department of Energy at Los Alamos National Lab as well as the US Department of Defense at the Defense Intelligence Agency as an applied mathematician of machine learning. He is at present working for the US Department of Energy at Sandia National Lab. His research interests include machine learning, statistics, and computational algebra.

A Beginner's Guide to Mathematical Proof

Mark J. DeBonis

CRC Press
Taylor & Francis Group
Boca Raton London New York

CRC Press is an imprint of the
Taylor & Francis Group, an **informa** business
A CHAPMAN & HALL BOOK

Designed cover image: Mark J. DeBonis

First edition published 2025
by CRC Press
2385 NW Executive Center Drive, Suite 320, Boca Raton FL 33431

and by CRC Press
4 Park Square, Milton Park, Abingdon, Oxon, OX14 4RN

CRC Press is an imprint of Taylor & Francis Group, LLC

ISBN: 978-1-032-68619-6 (hbk)
ISBN: 978-1-032-68770-4 (pbk)
ISBN: 978-1-032-68772-8 (ebk)

DOI: 10.1201/9781032687728

Publisher's note: This book has been prepared from camera-ready copy provided by the authors.

To Sebastian and Penelope

Friends for life

Contents

Preface   ix

CHAPTER   1 ▪ Mathematical Logic   1

1.1   PROPOSITIONAL LOGIC   1
1.2   THE IMPLICATION STATEMENT AND MATHEMATICAL
      PROOF   9
1.3   PREDICATE LOGIC   12

CHAPTER   2 ▪ Methods of Proof   17

2.1   DIRECT PROOF   17
2.2   SET THEORY   21
      2.2.1   Set Arithmetic   25
2.3   CONTRAPOSITIVE PROOF   30
2.4   PROOF BY CONTRADICTION   32
2.5   PROOF BY MATHEMATICAL INDUCTION   36
      2.5.1   Misuses of Mathematical Induction   45
2.6   PROOF BY CASES   48
2.7   REVIEW OF PROOF METHODS   56

CHAPTER   3 ▪ Special Proof Types   58

3.1   EXISTENCE PROOFS   58
      3.1.1   Non-Constructive Proofs   59
      3.1.2   Constructive Proofs   61

3.2   UNIQUENESS PROOFS   68

CHAPTER   4 ▪ Foundational Mathematical Topics   71

4.1   SET RELATIONS   71
        4.1.1   Cartesian Product   72
        4.1.2   Relation   79
        4.1.3   Equivalence Relation   83
4.2   FUNCTIONS   89
        4.2.1   Well-Definition   99
4.3   BASIC NUMBER THEORY   103
4.4   MODULO ARITHMETIC   106
4.5   SIZES OF INFINITY   110
        4.5.1   Countable versus Uncountable   110
        4.5.2   Equinumerous Sets   115
        4.5.3   Infinity Unbounded   119
4.6   SYMMETRIES AND COMBINATORICS   123
        4.6.1   Permutations   123
        4.6.2   Action   136
        4.6.3   Burnside's Lemma   141
        4.6.4   Polya's Formula   146

References   155

Index   157

Preface

This was perhaps one of my favorite courses to teach as a professor. It's the first class filled with mathematics majors and it's a chance to have a huge and positive impact on them. For some students demonstrating proofs comes naturally, while for others it requires a lot of hand holding. I wrote this text for the full spectrum of mathematics majors, and also for those who are just curious about how to demonstrate mathematical proof or want to understand how mathematical theory is developed.

There are no assumptions made about background for this text, so essentially anyone can pick this text up and begin reading and understanding. The goal of this text is not only to prepare the mathematics major for abstract mathematics, but also to give any reader an appreciation for it; to understand that applied mathematics has to come from somewhere, and it needs to be developed before it can get used. I believe to truly understand what mathematics is, one must understand theoretical mathematics. I find it amazing that people on the outside believe that mathematics is all about computation, as if mathematicians are simply very proficient calculators! But mathematics is so much more. It is both a beautiful and a creative subject, and mathematicians are a curious bunch full of extraordinary personalities.

This text should be easy to utilize by the instructor and equally easy to follow by the student, since it makes the most minimal of mathematical assumptions and was purposefully written to not include too many tangential or extraneous topics. The first two chapters are fundamental and no section should be skipped. Chapter three is less essential, although still very useful. Finally, chapter four was written to test your mettle once you've gone through the tools for performing mathematical proof. The first four sections in chapter four are foundational topics a mathematics major will see in studying theoretical mathematics, while the last two sections are meant to challenge the reader who wishes to take things to the next level.

My hope is that this text will be a useful addition to the list of texts already in print on this topic. I wish the reader good luck on your first steps in this field, and I hope you will come out with a deeper appreciation for what mathematics is actually about!

Special thanks to Aaron Powell for taking the time to read through this manuscript and make helpful editorial comments.

CHAPTER 1

Mathematical Logic

I N THIS CHAPTER we introduce the reader to the language and validity
of mathematics. We begin by introducing propositional logic and the
notion of truth tables. Of most importance is the implication, as we shall
see further along in the text. We then beef up the language to first-order
predicate logic, which is necessary for expressing mathematical ideas.

1.1 PROPOSITIONAL LOGIC

The reader may have encountered the material for this section in an
introduction to philosophy course.

Definition 1.1 *A* **statement** *(or* **proposition***) P is a phrase which
can be assigned a value of True or False but not both.*

Example 1.1 *Here we list a number of examples illustrating the definition of statement.*

1. *"One plus one equals two" is a True statement*

2. *$1 + 1 = 3$ is a False statement*

3. *"Water is wet" is a True statement*

4. *"The blue sky" is* **not** *a statement*

5. *"This sentence is a False statement" is not a statement (try to
   assign it a True or False value).*

6. *"New York City is the best city in the world" is* **not** *a statement,
   but rather an opinion.*

DOI: 10.1201/9781032687728-1

Definition 1.2 *A* **logical connective** *is a symbol used to modify a statement or to create a compound statement. These are*

> *The symbol* ¬ *which represents* **not**.

> *The symbol* ∧ *which represents* **and**.

> *The symbol* ∨ *which represents* **or**.

> *The symbol* → *which represents* **implication**.

> *The symbol* ↔ *which represents* **equivalence**.

> *The symbol* ⊕ *which represents* **exclusive or**.

Example 1.2 *Using the statements in Example 1.1, we create some compound statements.*

1. $1 + 1 = 3 \land$ *"One plus one equals two"*

2. *"Water is wet"* → *"Unicorns are real"*

Definition 1.3 *A* **truth table** *lists the possible truth values of a modified and/or compound statement based on the possible values of the statements included therein.*

Let's first look at the truth tables associated with the logical connectives already defined. Statements will be represented by letters P, Q, R, etc. The statement values True and False will be abbreviated by T and F, respectively. First, the truth table for **and**.

P	Q	$P \land Q$
T	T	T
T	F	F
F	T	F
F	F	F

In summary, the logical connective **and** is True exactly when both statements in the compound statement are True. Another way we can represent the truth table for **and** is as follows:

P	$\land$	Q
T	T	T
T	F	F
F	F	T
F	F	F

This second representation will become useful shortly when we look at the truth table for arbitrary compound statements. Now we present both representations of the **or** truth table.

P	Q	$P \vee Q$
T	T	T
T	F	T
F	T	T
F	F	F

P	$\vee$	Q
T	T	T
T	T	F
F	T	T
F	F	F

In summary, the logical connective **or** is True when at least one of the statements in the compound statement is True. This is in contrast to the **exclusive or** logical connective we now present.

P	Q	$P \oplus Q$
T	T	F
T	F	T
F	T	T
F	F	F

P	$\oplus$	Q
T	F	T
T	T	F
F	T	T
F	F	F

Indeed, the **exclusive or** is True when exactly one of the statements in the compound statement is True. The logical connective **implication** (or **conditional**) has the following truth table.

P	Q	$P \to Q$
T	T	T
T	F	F
F	T	T
F	F	T

P	$\to$	Q
T	T	T
T	F	F
F	T	T
F	T	F

Thus, the implication is False only in one instance of the truth table, namely "True implies False". Intuitively, if we begin with something we know to be True, we shouldn't be able to deduce something False.

There is terminology for the P and Q in the implication compound statement. P is referred to as the **hypothesis**, the **premise** or the **antecedent**. Q is referred to as the **conclusion** or the **consequent**.

Finally, we present the truth table for **equivalence**.

P	Q	$P \leftrightarrow Q$
T	T	T
T	F	F
F	T	F
F	F	T

P	$\leftrightarrow$	Q
T	T	T
T	F	F
F	F	T
F	T	F

Thus, equivalence is True exactly when the premise and conclusion have the same assigned value – either both True or both False.

Now with these symbols and these basic truth tables, one can derive the truth table for any compound statement. We take a naive approach when producing compound statements; however, the construction of valid compound statements can be formalized if one should take a course in mathematical logic [2].

Example 1.3 *Let's derive the truth table for several example compound statements. We will give a careful step-by-step derivation for these examples. In each example, P and Q are assumed to be statements.*

1. *Consider the compound statement $\neg P \vee Q$. We start by filling out all possibilities for the statements (which will create a table of length necessarily a power of two), and then fill in the blank columns in a particular logical order.*

$\neg$	P	$\vee$	Q
	T		T
	T		F
	F		T
	F		F

$\Rightarrow$

$\neg$	P	$\vee$	Q
F	T		T
F	T		F
T	F		T
T	F		F

$\Rightarrow$

$\neg$	P	$\vee$	Q
F	T	**T**	T
F	T	**F**	F
T	F	**T**	T
T	F	**T**	F

The final assignment of truth values to the compound statement is indicated in bold font.

2. Consider the compound statement ¬(¬P ∨ ¬Q).

¬	(¬	P	∨	¬	Q)
		T			T
		T			F
		F			T
		F			F

⇒

¬	(¬	P	∨	¬	Q)
	F	T		F	T
	F	T		T	F
	T	F		F	T
	T	F		T	F

⇒

¬	(¬	P	∨	¬	Q)
	F	T	F	F	T
	F	T	T	T	F
	T	F	T	F	T
	T	F	T	T	F

⇒

¬	(¬	P	∨	¬	Q)
T	F	T	F	F	T
F	F	T	T	T	F
F	T	F	T	F	T
F	T	F	T	T	F

3. Consider the compound statement (P → Q) ∧ (Q → P).

(P	→	Q)	∧	(Q	→	P)
T		T		T		T
T		F		F		T
F		T		T		F
F		F		F		F

⇒

(P	→	Q)	∧	(Q	→	P)
T	T	T		T	T	T
T	F	F		F	T	T
F	T	T		T	F	F
F	T	F		F	T	F

⇒

(P	→	Q)	∧	(Q	→	P)
T	T	T	**T**	T	T	T
T	F	F	**F**	F	T	T
F	T	T	**F**	T	F	F
F	T	F	**T**	F	T	F

4. Consider the compound statement (P ∨ Q) ∧ ¬(P ∧ Q).

(P	∨	Q)	∧	¬	(P	∧	Q)
T		T			T		T
T		F			F		T
F		T			T		F
F		F			F		F

⇒

(P	∨	Q)	∧	¬	(P	∧	Q)
T	T	T			T	T	T
T	T	F			F	F	T
F	T	T			T	F	F
F	F	F			F	F	F

(P	∨	Q)	∧	¬	(P	∧	Q)
T	T	T		F	T	T	T
T	T	F		T	T	F	F
F	T	T		T	F	F	T
F	F	F		T	F	F	F

$\Rightarrow$

(P	∨	Q)	∧	¬	(P	∧	Q)
T	T	T	**F**	F	T	T	T
T	T	F	**T**	T	T	F	F
F	T	T	**T**	T	F	F	T
F	F	F	**F**	T	F	F	F

Remark 1.1 *Several remarks are in order.*

1. *There is a natural order of operations when deciding which column in the truth table to fill in next. For instance, what is in parentheses must be filled out first. Also, if there is a negation, fill this in prior to evaluating a logical connective that takes this negation as one of its arguments.*

2. *Keep in mind that it is not required to write out all the partial truth tables you see presented in these examples; only the final truth table in which all the columns are filled, of course in the appropriate order.*

Definition 1.4 *A statement is called a*

1. **tautology** *if all the resulting values in its truth table are True.*

2. **contradiction** *if all the resulting values in its truth table are False.*

Example 1.4 *We will give one example of each.*

1. *$((P \to Q) \wedge P) \to Q$ is a tautology. Indeed, this statement is called* **modus ponens** *and is easy to interpret. It says that "if P implies Q and we have P, then Q follows". We will be using this implicitly when we do mathematical proof later on in the text. In other words,*

> *While demonstrating a mathematical proof, if you are assuming statement P implies statement Q is True and you are also assuming that statement P is True, then it follows that statement Q must also be True.*

2. *$P \wedge \neg P$ is a contradiction, which makes sense since we cannot have both a statement and its opposite be True at the same time.*

Definition 1.5 *Two statements P and Q are* **logically equivalent**, *written $P \equiv Q$, if they have the same truth tables.*

One can think of logically equivalent statements as essentially equality, since in all cases logically equivalent statements are True or False exactly at the same time. In other words,

> While demonstrating a mathematical proof, if two statements are logically equivalent and you are assuming one of those statements are True, then it follows that the other statement must also be True.

Example 1.5 *Refer to the truth tables in Example 1.3 to convince yourself of the logical equivalences listed below for arbitrary statements P and Q.*

1. $P \rightarrow Q \equiv \neg P \vee Q$.

2. $P \wedge Q \equiv \neg(\neg P \vee \neg Q)$.

3. $P \leftrightarrow Q \equiv (P \rightarrow Q) \wedge (Q \rightarrow P)$.

4. $P \oplus Q \equiv (P \vee Q) \wedge \neg(P \wedge Q)$.

Remark 1.2 *We wish to make several remarks about logical equivalence.*

1. *Example 1.5.4, for instance, is easy to interpret. It says that exclusive or is True when P or Q is True but not both.*

2. *Examples 1.5.1-.4 tell us that we only need $\neg$ and $\vee$ to express compound statements, and the rest of the logical connectives can be viewed as abbreviations.*

Example 1.6 *We present some additional logical equivalent facts, but we leave the verification to the reader.*

1. $\neg(\neg P) \equiv P$.

2. $\neg(P \wedge Q) \equiv \neg P \vee \neg Q$.

3. $\neg(P \vee Q) \equiv \neg P \wedge \neg Q$.

4. *Assuming these facts and those in Example 1.5 and 1.6.1-.3, we verify another logical equivalence without the need for truth tables. We will show that $\neg(P \rightarrow Q) \equiv P \wedge \neg Q$.*

$$\neg(P \rightarrow Q) \equiv \neg(\neg P \vee Q) \equiv \neg(\neg P) \wedge \neg Q \equiv P \wedge \neg Q.$$

EXERCISES

1 Fill out the truth table for each of the following statements, where P, Q, and R are statements.

   a. $(P \wedge Q) \rightarrow P$.

   b. $P \leftrightarrow (Q \vee R)$.

2 Verify that

   a. $P \rightarrow (P \vee Q)$ is a tautology,

   b. $((P \rightarrow Q) \wedge P) \rightarrow Q$ is a tautology, and

   c. $P \wedge \neg P$ is a contradiction.

3 Verify logical equivalence for each exercise below, where P, Q, and R are statements.

   a. $\neg(\neg P) \equiv P$.

   b. $P \rightarrow Q \equiv \neg Q \rightarrow \neg P$. This latter statement is call the **contrapositive** statement for the implication.

4 Verify logical equivalence for each exercise below, where P, Q, and R are statements. These exercises illustrate an associativity and distributive property for $\wedge$ and $\vee$.

   a. $P \vee (Q \vee R) \equiv (P \vee Q) \vee R$.

   b. $P \wedge (Q \wedge R) \equiv (P \wedge Q) \wedge R$.

   c. $P \wedge (Q \wedge R) \equiv (P \wedge Q) \wedge (P \wedge R)$.

   d. $P \wedge (Q \vee R) \equiv (P \wedge Q) \vee (P \wedge R)$.

   e. $(P \vee Q) \wedge R \equiv (P \wedge R) \vee (Q \wedge R)$.

5 Verify the logical equivalence $(P \vee Q) \wedge (P \rightarrow Q) \equiv Q$, where P and Q are statements.

6 Verify DeMorgan's Laws, namely

   a. $\neg(P \wedge Q) \equiv \neg P \vee \neg Q$, and

   b. $\neg(P \vee Q) \equiv \neg P \wedge \neg Q$

7 Decide whether or not each of the following statements are True:

    a. $[(P \to Q) \wedge (\neg P \to Q)] \equiv Q$

    b. $[(P \wedge Q) \vee (\neg P \wedge Q)] \equiv P \leftrightarrow Q$

8 Verify the following using a partial truth table.

    a. If statement P is known to be True, then $P \wedge Q$ is logically equivalent to Q.

    b. If statement P is known to be False, then $P \vee Q$ is logically equivalent to Q.

9 Using facts, examples and exercises already presented and without using truth tables, verify the following logical equivalence:

$$(P \vee Q) \to R \equiv (P \to R) \wedge (Q \to R).$$

1.2 THE IMPLICATION STATEMENT AND MATHEMATICAL PROOF

We wish to point out that another way to say $P \to Q$, i.e. "P implies Q" is to say "If P, then Q". As we shall see later on in the text, any mathematical statement we wish to prove to be True can be phrased as an implication. Indeed, this is the approach we wish to take when we engage in mathematical proof.

Take, for instance, the statement "The sum of two odd integers is an even integer". Another way we can phrase this statement is as follows: "If two integers are odd, then their sum is even"; hence, we can rephrase the original statement as an implication. For this reason in this section we shall take a closer look at the implication statement.

First note that if we wish to prove an implication is True we may as well assume the premise is True. Indeed, look again at the truth table for implication and you shall see that in all cases when the premise is False – whether the conclusion is True or False – the implication is always True. So there is no need to consider the case where the premise is False, since the implication is always True in this situation. We say in this case that the implication **holds vacuously**. Therefore,

When mathematically proving that an implication $P \to Q$ is True, we will assume the premise is True and through a series of logical steps show that the conclusion is also True.

There are four special implications we wish to consider.

Definition 1.6 *Let P and Q be any statements.*

1. $P \to Q$ *is sometimes call the* **direct** *statement.*

2. $\neg Q \to \neg P$ *is called the* **contrapositive** *statement.*

3. $Q \to P$ *is called the* **converse** *statement.*

4. $\neg P \to \neg Q$ *is called the* **inverse** *statement.*

Remark 1.3 *We will make several remarks about these four types of implications and how they relate to each other.*

1. *As we saw in Exercise 3b. of Section 1.1, the contrapositive statement is logically equivalent to the direct statement. Informally, to see this, $P \to Q$ says whenever you have P you also have Q, while the contrapositive says if you didn't have Q, then you couldn't have had P. These statements are semantically equivalent.*

2. *The converse statement is* **not** *logically equivalent to the direct conditional. One can make this evident by computing the truth table for each and observing that they are not the same, but let's get an intuitive understanding with a simple example. Let P be the statement "x is an integer" and Q be "x^2 is an integer". Certainly P implies Q is always True for any value of x, however this is not the case for the converse. Just consider the case $x = \sqrt{2}$. The statement Q is True, however P is False, and thus $Q \to P$ is False.*

3. *The inverse statement is logically equivalent to the converse statement. Indeed, the inverse is the contrapositive of the converse and therefore they are logically equivalent.*

There is additional language when discussing the implication which we present here.

Definition 1.7 *Let P and Q be arbitrary statements and consider the implication $P \to Q$. We say*

1. Q is **necessary** for P, and

2. P is **sufficient** for Q.

Remark 1.4 *Mathematicians often use this terminology and it can be sometimes confusing, so let's understand it in an intuitive way.*

1. *The statement "Q is **necessary** for P" is saying that if you have P you necessarily have Q. Let's refer back to the example in Remark 1.3.2. If P is True, i.e. x is an integer, then Q is necessarily True. Indeed, in this case x^2 is an integer.*

2. *The statement "P is **sufficient** for Q" is saying to have Q it is sufficient to have P (but not necessary). Again, referring to Remark 1.3.2, the statement is saying it is sufficient for x to be an integer in order for x^2 to be an integer, but it's not necessary, since $x = \sqrt{2}$ (which is not an integer) makes Q an integer.*

3. *By Example 1.5.3 of Section 1.1 $P \leftrightarrow Q$ is logically equivalent to $(P \to Q) \wedge (Q \to P)$, and so P is necessary and sufficient for Q. Indeed, in this case mathematicians use the terminology P iff Q, where "iff" stands for "if and only if", or in other words P "exactly when" Q.*

4. *This last remark tells us that one approach for giving a mathematical proof for $P \leftrightarrow Q$ is to demonstrate both $P \to Q$ and $Q \to P$, once again illustrating the fact that mathematical proof is all about verifying implication statements are True.*

EXERCISES

1 Rephrase each of the statements below as an "if/then" statement.

    a. "The sum of two even integers is even".

    b. "The product of a even integer and any other integer is even".

2 For each statement in Exercise 1, express in words the contrapositive, converse and inverse statements.

3 Create a True direct statement for which the converse is also True.

4 Create a True direct statement for which the converse is False.

1.3   PREDICATE LOGIC

Propositional Logic is not expressive enough to represent all of mathematics. We need a more extensive language. We need some additional symbols, namely variables, quantifiers, equality and membership. We could make this presentation much more formal, but as an introductory text we will once again take a more naive approach. For the reader more interested in the subject, a course in mathematical logic would serve this purpose [2].

Definition 1.8 *A* **variable** *represents an unspecified object and may be represented by an arabic letter, for instance x, or in a list of subscripted letters, say $x_1, x_2, \ldots, x_n$.*

A **sentence** $P(x_1, x_2, \ldots, x_n)$ *is a phrase which includes the variables* $x_1, x_2, \ldots, x_n$.

Example 1.7 *An example of a sentence $P(x_1, x_2)$ might be*

$$(x_1 + x_2 = 4) \wedge (x_1 \cdot x_2 = 2).$$

One could interpret this as saying "The sum of two objects is four and the product is two". Notice that this sentence is neither True nor False, since the variables are simply place holder objects. Indeed this is the case for any sentence. Notice also that we have introduced two binary operators $+$ and $\cdot$ (as well as two constants 4 and 2) in a very naive sort of way, relying on the reader's familiarity with these symbols, so we could interpret these objects as numbers of a sort (again, this could be made more formal).

Definition 1.9 *There are two quantifiers in predicate logic.*

1. *The* **universal** *quantifier, denoted by $\forall$, and is read as "For every" or "For all".*

2. *The* **existential** *quantifier, denoted by $\exists$, and is read as "There exists" or "For some".*

Example 1.8 *We will apply some quantifiers to Example 1.7 to obtain*

$$\forall x_1 \exists x_2 \; (x_1 + x_2 = 4) \wedge (x_1 \cdot x_2 = 2).$$

One could interpret this as saying "For every number there exists another number such that the sum of these numbers is four and their *product is two". It is still difficult to decide if this statement is True or False, since we need more context. Indeed, we are interpreting x_1 and x_2 as numbers because of the use of addition and multiplication, but the type of numbers we are referring to is still unknown.*

Hence, for the sake of this discussion we quickly introduce some number systems.

1. The **natural** numbers, denoted by $\mathbb{N}$, is the list of numbers $0, 1, 2, 3, \ldots$, etc.

2. The **integers**, denoted by $\mathbb{Z}$, is the list of numbers $0, \pm 1, \pm 2, \pm 3, \ldots$, etc.

3. The **rational** numbers, denoted by $\mathbb{Q}$, is the collection of all fractions of integers p/q where q is non-zero.

4. The **real** numbers, denoted by $\mathbb{R}$, is the collection of all decimal expansions, e.g. $\pi = 3.1415926\cdots$ or $1/16 = 0.0625$.

5. The **irrational** numbers are real numbers which are **not** rational numbers, e.g. $\pi = 3.1415926\cdots$.

Definition 1.10 *The symbol $\in$ will be the* **membership function** *and can be read as "an element of" or simply as "in".*

Example 1.9 *We will supply some context to Example 1.8 to obtain*

$$\forall x_1 \in \mathbb{R} \; \exists x_2 \in \mathbb{R} \; (x_1 + x_2 = 4) \wedge (x_1 x_2 = 2).$$

Now one could interpret this as saying "For every real number there exists another real number such that the sum of these numbers is four and the product is two". Now we can decide if this (now a statement) is True or False. Indeed, if all variables are quantified and given context, then we can assign a True or False value.

We will show that the statement is False. We illustrate this by way of a **counterexample***, which is an example that demonstrating a statement is False. If we can find an example of a real number for which there is no other real number to add to it to make four and multiply by it to make two, then we have succeeded. Consider the number $x_1 = 1$. Since the sum is four, it must be the case that $x_2 = 3$; however, $(1)(3) = 3 \neq 2$,* *thus we have found a counterexample. Indeed, there are an infinite number of counterexamples (for instance, any integer will work as a counterexample for this example).*

Now consider the slightly different statement

$$\exists x_1 \in \mathbb{R} \; \exists x_2 \in \mathbb{R} \; (x_1 + x_2 = 4) \wedge (x_1 x_2 = 2).$$

This statement says there exist two real numbers whose sum is four and product is two. We will show this is True by finding such a pair of real numbers. This is just a problem in basic algebra. Indeed, $x_2 = 4 - x_1$ and so $x_1(4 - x_1) = 2$ or $x_1^2 - 4x_1 + 2 = 0$. Then using the quadratic formula

$$x_1 = (4 \pm \sqrt{8})/2 = 2 \pm \sqrt{2}.$$

If $x_1 = 2 + \sqrt{2}$, then $x_2 = 4 - x_1 = 2 - \sqrt{2}$. Notice that, $x_1 + x_2 = 4$ and $x_1 x_2 = (2 + \sqrt{2})(2 - \sqrt{2}) = 4 - 2 = 2$. Indeed, this is the only pair of numbers making the statement True.

Example 1.10 *The purpose of this example is to illustrate that the order of quantifiers is important. Consider the following two statements in which the only difference is the order of the quantifiers.*

1. $\forall x \in \mathbb{R} \; \exists y \in \mathbb{R} \; x + y = 0.$

2. $\exists y \in \mathbb{R} \; \forall x \in \mathbb{R} \; x + y = 0.$

The first statement is True. Indeed, the y we seek is $-x$. However, the second is certainly False for no matter what real number y should be, there is no way to add to every real number and always get 0.

We state the following facts which should be self-evident. These deal with the negation of quantified statements. Let $P(x)$ be a sentence in the variable x. Then

1. $\neg(\forall x \; P(x)) \equiv \exists x \; \neg P(x)$

2. $\neg(\exists x \; P(x)) \equiv \forall x \; \neg P(x)$

Indeed, the first statement says that if it is not the case $P(x)$ for all x, then there must be some x for which $P(x)$ is not the case. The second statement says that if it is not the case $P(x)$ for some x, then $\neg P(x)$ must be the case for all x. Note that the context of these variables will not change the validity of these facts.

Example 1.11 *We shall negate the following statements:*

1. $\forall x \in \mathbb{R} \ x > 0$

2. $\forall x \in \mathbb{R} \ \exists y \in \mathbb{R} \ x = y^2.$

For the first statement,

$$\neg(\forall x \in \mathbb{R} \ x > 0) \equiv \exists x \in \mathbb{R} \ \neg(x > 0) \quad \textit{or} \quad \exists x \in \mathbb{R} \ x \leq 0.$$

One way to interpret the first statement is to say if it is not the case that every real number is positive, then there must be some real number which is either zero or negative. For the second statement,

$$\neg(\forall x \in \mathbb{R} \ \exists y \in \mathbb{R} \ x = y^2) \equiv \exists x \in \mathbb{R} \ \neg(\exists y \in \mathbb{R} \ x = y^2)$$

$$\equiv \exists x \in \mathbb{R} \ \forall y \in \mathbb{R} \ x \neq y^2.$$

We leave it to the reader to interpret this second negated statement.

This technique for negating quantified statements will become useful in our methods of proof as we shall see later on the text (cf. *proof by contrapositive* or *proof by contradiction*).

One final note before we complete this section and the chapter is to present some useful abbreviations for the existence quantifier which are sometimes used by mathematicians.

1. $\exists! x \ P(x)$ is to be interpreted as "There exists a **unique** x such that $P(x)$". Indeed, we can write this in the logically equivalent way without any abbreviation as

$$\exists x \ P(x) \wedge \forall y \ P(y) \rightarrow y = x.$$

2. $\exists^{\leq n} x \ P(x)$ is to be interpreted as "There exists **no more than** n different x such that $P(x)$".

EXERCISES

1 Decide whether or not each of the following statements are True or False. If False, provide a counterexample.

a. $\forall x \in \mathbb{R} \ x > 0.$

b. $\exists x \in \mathbb{R} \ x > 0.$

   c. $\exists x \in \mathbb{R} \; x^2 < 0$.

   d. $\forall x \in \mathbb{R} \; \exists y \in \mathbb{R} \; x = y^2$.

   e. $\forall x \in \mathbb{N} \; \exists y \in \mathbb{R} \; x = y^2$.

   f. $\forall x \in \mathbb{N} \; \exists y \in \mathbb{N} \; x = y^2$.

   g. $\exists x \in \mathbb{R} \; \exists y \in \mathbb{R} \; x = y^2$.

   h. $\forall x \in \mathbb{R} \; x > 0 \rightarrow \exists y \in \mathbb{R} \; x = y^2$.

2 In Example 1.11.2, write in words an interpretation of the negation of the statement and its resulting logically equivalent statement.

3 Write the following mathematical statement as a statement in words (i.e. with no symbols):

$$\forall x \in \mathbb{R} \; \exists y \in \mathbb{N} \; (x = y) \vee (x < y).$$

4 Negate (and simplify) the mathematical statement in the previous exercise.

5 Write the following statement in words as a mathematical statement (i.e. symbols only):

"Any two real numbers are either equal, or one is less than the other, or vice versa".

6 Negate the following statements:

   a. $\exists x \in \mathbb{R} \; x^2 < 0$.

   b. $\forall x \in \mathbb{R} \; x < 0 \rightarrow \exists y \in \mathbb{R} \; x = y^2$.

   Hint: convert the implication to a logically equivalent statement involving $\vee$.

   c. $\forall x \in \mathbb{R} \; [\exists y \in \mathbb{Z} \; (y > x) \leftrightarrow \forall z \; (x = z)]$

7 Express $\exists^{\leq n} x \; P(x)$ in a logically equivalent way without using abbreviation.

CHAPTER 2

Methods of Proof

I N THIS CHAPTER we introduce the reader to the full toolbox of proof methods needed by the mathematician. This includes the direct proof, contrapositive proof, proof by contradiction, proof by induction and proof by cases. In doing so we need topics upon which to employ our methods. Therefore, we also introduce basic notions in set theory and number theory.

2.1 DIRECT PROOF

As we have already discussed in Section 1.2, in order to prove a mathematical statement, we first phrase it as an implication $P \to Q$. We then assume P is True and through a series of steps conclude that Q is also True. The steps by which we arrive at the conclusion is the creative part of mathematical proof [5]. It is not something that can be given as a recipe; it is not something you can express as a computer algorithm, although there has been some inroads made by AI in recent years. Nevertheless, there are some rules of thumb one can keep in mind while creating a sound mathematical proof. These we will point out as we progress through this chapter.

Recall the mathematical statement "The sum of two odd integers is an even integer" which we rewrote as an implication "If two integers are odd, then their sum is even". In proving this implication we may assume the premise that we have two odd integers, say m and n. Our goal is to show that $m + n$ is even. One rule of thumb in mathematical proof is to

Express terminology in terms of formal definitions.

So let's give a formal definition of even and odd.

DOI: 10.1201/9781032687728-2

Definition 2.1 *An integer is*

1. **even** *if it can be written as $2k$ for some integer k.*

2. **odd** *if it can be written as $2k + 1$ for some integer k.*

Continuing with our proof, since m and n are odd we know $m = 2k+1$ for some integer k and $n = 2l+1$ for some integer l. Notice that we choose a new variable l in reference to n, otherwise we have $m = 2k + 1 = n$, which spoils the generality of the proof. We never assumed that m and n are equal in the premise of the implication. This is another rule of thumb to keep in mind as you demonstrate a proof, namely

---
Always ask yourself if you are making any additional assumptions not given as you step through your proof.

Now consider the sum

$$m + n = (2k + 1) + (2l + 1).$$

It is not yet clear that $m + n$ is an even integer, but perhaps we can rewrite the sum in such a way that it is clear. Notice

$$m + n = 2k + 1 + 2l + 1 = 2k + 2l + 2 = 2(k + l + 1).$$

Hence, we expressed the sum as two times an integer and it is therefore even. This completes the proof.

Notice that we are making an assumption that $k+l+1$ is an integer. This is a basic assumption about integers that when you add them together you again get an integer. So at a certain point we have to accept certain results as facts (although, even this can be formally proved – an introductory text on set theory would cover this topic [3]).

Example 2.1 *Let's prove another mathematical statement, namely "If a, b, and c are real numbers with $a > b$ and c positive, then $ac > bc$". Before we can do so, we need to formally define $>$.*

Definition 2.2 *Let a and b be any two real numbers. We say a is* **greater than** *b, written $a > b$, if the difference $a - b$ is positive.*

Now we can prove the result.

Proof 2.1 *Assume that a, b, and c are integers with a > b and c positive. By definition of >, we know that a − b is positive. But then (a − b)c is also positive. Another way to write this – using the distributive rule – is ac − bc is positive. But then again, by definition of greater than, we have ac > bc, and our proof is complete.*

Again, we made some basic assumptions about positive numbers, namely that the product of two positive numbers is again positive. There will always be some basic things we will have to assume which could be proved formally, but would take us too far afield. There is a certain knack for deciding when this is the case, and it can be frustrating at first to the mathematical initiate.

Example 2.2 *Let's prove one more mathematical statement, namely "If a, b, and c are integers with a divides b and b divides c, then a divides c". To say one integer a divides another integer b means that a divides evenly into b or that a is a factor of b (for instance, 3 divides 15). We now formally define what we mean by divides.*

Definition 2.3 *Given two integers a and b, we say a **divides** b and we write a | b, if there exists an integer k such that b = ak.*

As an example $-3 \mid 15$, since $15 = (-3)(-5)$ (in this example $k = -5$). Now we can prove the result.

Proof 2.2 *Assume a, b and c are integers with a divides b and b divides c. By definition of divides, there exist integers k and l such that b = ak and c = bl (again, note that k and l may be different integers and have to be designated as such). Putting these two equations together,*

$$c = bl = (ak)l = a(kl).$$

Since kl is an integer, by definition of divides we have that a divides c, and the proof is complete.

Once again, we made some basic assumptions about integers. First, that they satisfy the associative rule for multiplication and second, that the product of two integers is again an integer.

Remark 2.1 *There are several ways one can indicate to the reader that your proof is complete. Some may express it in words, as was done in the* *proof above. Others like to end their proof in QED which stands for "quod erat demonstrandum" which is Latin for "which was to be demonstrated". Others use four diagonal lines, ////, to end their proof. In this text we will use the symbol □ to indicate the proof is complete. One might ask why a square? As the story goes, the use of the square to represent the end of a proof was first introduced by the mathematician Paul Halmos.*

EXERCISES

1 Give a counterexample for the following statement:

$$\forall a, b, c \in \mathbb{Z} \ (\ a \mid c \ \wedge \ b \mid c \ ) \ \rightarrow \ (ab) \mid c.$$

2 Use a direct proof to demonstrate the following results about integers.

   a. The sum of two even integers is even.

   b. The sum of an even and odd integer is odd.

   c. The product of two odd integers is odd.

   d. An even integer times any integer is even.

3 Prove that if the square root of a positive real number is rational, then the number itself must be rational.

4 Use a direct proof to demonstrate the following results about inequalities.

   a. If a, b, c, and d are real numbers with $a > b$ and $c > d$, then $a + c > b + d$.

   b. If a, b are real numbers with $a > b \geq 0$, then $a^2 > b^2$.

   c. If a, b are real numbers with $a > b \geq 0$, then $\sqrt{a} > \sqrt{b}$.

   d. If a, b are real numbers with $a > b \geq 0$, then $1/b > 1/a$.

5 Use a direct proof to demonstrate the following results about divides.

   a. If a, b, and c are integers with $a \mid b$, then $a \mid (bc)$.

   b. If a, b, m, n, and d are integers with $d \mid m$ and $d \mid n$, then $d \mid (am + bn)$.

c. If m and n are integers with $m \mid n$ and $n \mid m$, then $m = \pm n$.

d. If m and n are positive integers with $m \mid n$, then $1 \leq m \leq n$.

2.2 SET THEORY

A fundamental component of mathematics is set theory. It is a wonderful subject in its own right and many mathematicians devote their time and research entirely in this subject alone. For us, we will look at some of the basic ideas and building blocks and present the minimal amount to carry on in this text.

Definition 2.4 *A* **set** *is a collection of objects. An object in the set is called an* **element** *of the set. A set is* **finite** *if it has a finite number of elements. A set is* **infinite** *if it has an infinite number of elements. The size of a finite set A, written $|A|$, equals the number of elements in the set A. If A is infinite, we write $|A| = \infty$.*

Example 2.3 *Sets are oftentimes represented by notation such as A, B, C, etc. Here are some examples.*

1. *$A = \{0, 1, 2, 3\}$, a finite set and $|A| = 4$.*

2. *$\mathbb{N} = \{0, 1, 2, 3, \ldots\}$, the natural numbers which is an infinite set; as are the integers $\mathbb{Z}$, the rationals $\mathbb{Q}$ and the reals $\mathbb{R}$.*

3. *$B = \{a, b, f\}$ is another finite set. So is $C = \{\triangle, \square, \diamond, \partial\}$.*

4. *If A, B, and C are sets, then so is $D = \{A, B, C\}$ a finite set of sets.*

5. *P can be the set of all finite sets of natural numbers. One such element of P is the set $\{99, 107, 1232, 10009\}$. Certainly P is an infinite set, since P contains the singleton sets $\{n\}$ for every integer $n \in \mathbb{Z}$ and $\mathbb{Z}$ is an infinite set.*

In the examples above we represent sets as lists of elements between two curly brackets. This type of representation of a set is sometime called **roster** notation. However, not all sets can be expressed in this way. We need a notation which is more versatile. The more general notation we will call **set builder** notation and has the form:

$$\{x \mid \mathcal{P}(x)\} \quad \text{or} \quad \{x : \mathcal{P}(x)\}, \quad \text{for some well-defined property } \mathcal{P}.$$

Example 2.4 *Here are some examples of sets defined using set builder notation.*

1. $A = \{x \mid x \in \mathbb{Z} \text{ and } x \geq 0\}$. *This would be read as "the set of all integers which are non-negative". What we've described in a more round about way is the natural numbers. As an alternative we can also write* $\{x \in \mathbb{Z} \mid x \geq 0\}$. *In other words we can state up front the context for our objects – in this case integers.*

2. *Although not impossible, it is not so easy to represent the rational numbers using roster notation. Using set builder notation*

$$\mathbb{Q} = \{x \mid x = p/q, \ p, q \in \mathbb{Z} \text{ and } q \neq 0\}$$

or $\{p/q \mid p, q \in \mathbb{Z} \text{ and } q \neq 0\}$.

3. *There is no way to represent real numbers in roster form. The technical reason for why this is the case is that the real numbers are not what mathematicians call* **countable***, terminology you would hear in a set theory course and which we will discuss later on in the text. We will take the cheap way out and define them as*

$$\mathbb{R} = \{x \ : \ x \ \text{has a decimal expansion}\} \ \text{or}$$

$\{x \mid x = d.d_1 d_2 \cdots \ \text{where} \ d \in \mathbb{Z} \ \text{and} \ d_1, d_2, \ldots \in \{0, 1, 2, \ldots, 9\}\}$.

4. $A = \{x \ : \ x \ \text{is nice}\}$ *is* **not** *a set, since "nice-ness" is not a well-defined property.*

5. *A well-known example of a collection of objects which do not form a set is often referred to as* **Russell's Paradox** *after the philosopher and mathematician, Bertrand Russell. Let* $B = \{A \mid A \notin A\}$. *One might ask how a set can be an element of itself, however there is an intuitive way to understand this example. Think of sets as catalogs, i.e. documents which list objects of a certain type such as the catalog of all dog breeds or the catalog of all Italian car models. Then* B *can be interpreted as the catalog of all catalogs which do not list themselves in their own catalog. The question then is does the catalog* B *list itself? Mathematically, we mean to say is it true that* $B \in B$, *or perhaps* $B \notin B$? *In either case one*

is lead to a contradiction – hence, the paradox. For if $B \in B$ then by definition of B we must conclude that $B \notin B$, a contradiction. On the other hand, if $B \notin B$, then by the definition of B it follows that $B \in B$, again a contradiction. Hence, B does not have a well-defined property and so cannot be a set.

Here are some additional definitions we will need for our discussion of set theory.

Definition 2.5 *Let A and B be sets.*

- *The **empty** or **null** set, written as $\emptyset$ or $\{\ \}$, is the set containing no elements.*

- *The **cardinality** or **size** of a set, written $|A|$, is the number of elements in the set. If A is an infinite set, we will write $|A| = \infty$.*

- *Two sets are **equal**, written $A = B$, if the two sets contain the exact same elements.*

- *A is a **subset** of B, written $A \subseteq B$, if every element of A is also in B.*

- *A is a **proper subset** of B, written $A \subset B$ or $A \subsetneq B$, if A is a subset of B yet is not equal to B. One says A is **properly contained** in B. In other words, $A \subseteq B \wedge \neg(A = B)$.*

Remark 2.2 *Here, we make several remarks about the definitions just presented.*

1. *The order in which we list the elements in a set does not change the set. For instance,*

$$\{0, 1, 2, 3\} = \{2, 0, 3, 1\}.$$

2. *It is pretty obvious when two finite sets have the same size, for instance*

$$A = \{0, 1, 2, 3\} \quad and \quad B = \{a, b, c, d\},$$

however, it is not so clear this is true when the sets are infinite in size. This was an important topic of study by the German mathematician Georg Cantor. He proved that infinite sets can have different sizes. Without going into detail for the moment, one can show that $|\mathbb{N}| = |\mathbb{Z}| = |\mathbb{Q}|$, however, $|\mathbb{N}| < |\mathbb{R}|$.

3. *The fact that $|\mathbb{N}| = |\mathbb{Q}|$ might be counterintuitive to most readers, since $\mathbb{N}$ is properly contained in $\mathbb{Q}$. Indeed, for any natural number $n \in \mathbb{N}$, we can express n as $n/1$ and, thus, it's an element of $\mathbb{Q}$. However, there are clearly many elements of the rationals which are not integers, like $3/4$. Indeed, there are an infinite number of elements in $\mathbb{Q}$ which are not natural numbers. Take, for instance, the infinite number of elements of the form $n/(n+1)$ for $n = 1, 2, 3, \ldots$, i.e. $1/2$, $2/3$, $3/4$, etc.*

4. *We make the obvious remark that not all sets are comparable as subsets of one another. For instance the sets $\{1, 3, 5\}$ and $\{0, 1, 3\}$. Indeed,*

$$\{1, 3, 5\} \nsubseteq \{0, 1, 3\} \quad and \quad \{0, 1, 3\} \nsubseteq \{1, 3, 5\}.$$

*And certainly, being a smaller set does not guarantee subset, for instance, although $\{1, 3\} \subsetneq \{1, 3, 5\}$ we have $\{0, 3\} \nsubseteq \{1, 3, 5\}$. For this reason we call subset a **partial ordering** of the collection of sets. This notion of parital ordering is important and will be made more explicit later on in the text.*

Now, in mathematical proof, when one wants to show subset or equality of sets we take the following approach.

- To prove that $A \subseteq B$, one needs to prove the statement "If $x \in A$, then $x \in B$". This is sometimes called **element chasing**.

- To prove $A = B$, show both $A \subseteq B$ and $B \subseteq A$, and to do this one needs to prove two statements.

  1. "If $x \in A$, then $x \in B$", and
  2. "If $x \in B$, then $x \in A$"

  Equivalently, show that "$x \in A$ iff $x \in B$".

- To prove $A \subsetneq B$, show $A \subseteq B$ and $\exists x \in B$ such that $x \notin A$.

- To prove $A \not\subseteq B$, i.e. A is not a subset of B, show
$\exists x \, (x \in A) \wedge (x \notin B)$.

Let's apply element chasing to prove the following results:

Theorem 2.1 *For any sets A, B, and C,*

1. $\emptyset \subseteq A$.

2. $A \subseteq A$.

3. *If $A \subseteq B$ and $B \subseteq C$, then $A \subseteq C$.*

Proof 2.3 *For the first statement, we need to show "If $x \in \emptyset$, then $x \in A$". But since $\emptyset$ is the empty set, the statement "$x \in \emptyset$" must be False, and therefore the implication "If $x \in \emptyset$, then $x \in A$" must be True, i.e. the staement holds vacuously.*

For the second statement, we need to show "If $x \in A$, then $x \in A$". Assuming as we do that the premise "$x \in A$" is True, since the conclusion is identical to the premise we therefore have that the conclusion is also True, thus making the implication also True.

For the third statement, as an implication we may assume that $A \subseteq B$ and $B \subseteq C$ are True and we need to show that $A \subseteq C$ is also True. To prove $A \subseteq C$, we need to show "If $x \in A$, then $x \in C$". Assuming the premise is True, i.e. "$x \in A$", since $A \subseteq B$, this implies $x \in B$ is also True. In addition, since $B \subseteq C$ is True, this implies in turn that $x \in C$ is True, and thus we have arrived at the conclusion. □

2.2.1 Set Arithmetic

There are certain operations you can apply to sets which we now introduce.

Definition 2.6 *Let A and B be sets.*

1. *The* **union** *of sets A and B, written*

$$A \cup B = \{x \mid x \in A \ \text{ or } \ x \in B\}.$$

In other words, the union of two sets consists of elements which are in either A or in B.

2. *The* **interection** *of sets* A *and* B*, written*

$$A \cap B = \{x \mid x \in A \quad and \quad x \in B\}.$$

In other words, the intersection of two sets consists of elements which are in both A *and* B.

3. *The set* **difference** *of* A *minus* B*, written*

$$A - B = \{x \mid x \in A \quad and \quad x \notin B\}.$$

Alternate notation for set difference is $A \backslash B$.

4. *The* **complement** *of set* A*, written*

$$A' = \{x \mid x \notin A\}.$$

Alternate notation for complement are $\overline{A}$ *or* A^c. *One needs some context to formally define the complement of a set. We need a set called the* **universal** *set or just* **universe***, and is typically denoted by* U. *It represents the set which contains all sets. The astute reader may notice that this borders again on Russell's paradox, however there is a way to avoid this paradox by always including in your set builder definition that each element in the set is also a set (but this goes beyond our naive treatment of set thoery, so we will leave it at that). The point is that*

$$A' = U - A.$$

5. *The* **symmetric difference** *of two sets* A *and* B*, written*

$$A \triangle B = (A - B) \cup (B - A).$$

It is called symmetric, since $A \triangle B = B \triangle A$.

6. *A union of two sets is called a* **disjoint union***, written* $A \sqcup B$ *if* $A \cap B = \emptyset$. *In other words, the sets* A *and* B *have no elements in common.*

Perhaps one way you have seen sets represented including their union, intersection, difference, etc. is with Venn diagrams (see Figure 2.1).

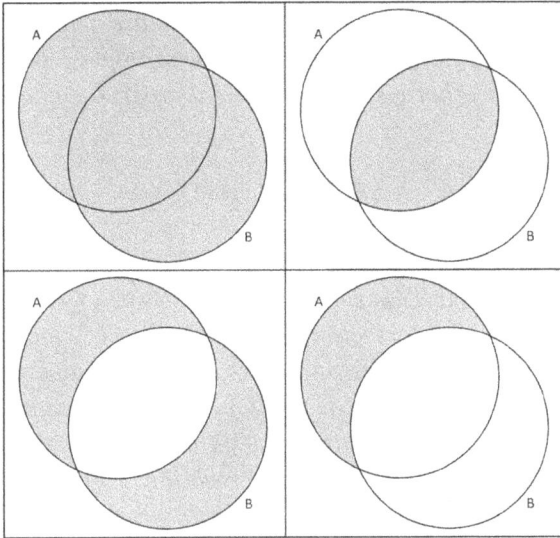

Figure 2.1 Starting in the upper left-hand corner and going clockwise, we have the Venn diagrams representing $A \cup B$, $A \cap B$, $A - B$, and $A \triangle B$.

Example 2.5 *For the sake of simplicity, let's suppose our universal set is the natural numbers, and consider the sets A of positive even numbers and B of positive multiples of five. In other words*

$$U = \mathbb{N}, \quad A = \{2, 4, 6, 8, \ldots\} \quad and \quad B = \{5, 10, 15, 20, \ldots\}.$$

Then

$$A \cup B = \{2, 5, 4, 6, 8, 10, 12, 14, 15, \ldots\} \quad and \quad A \cap B = \{10, 20, 30, 40, \ldots\}.$$

$$A - B = \{2, 4, 6, 8, 12, 14, 16, 18, 22, \ldots\} \quad and \quad A' = \{1, 3, 5, 7, \ldots\}.$$

Venn diagrams can be useful at times to visualize a mathematical statement about sets, however they are not general enough (and sometimes impossible to represent). In practice, they cannot take the place of a formal mathematical proof, which typically requires element chasing; however, it can sometimes lend some insight into whether or not the statement is True.

Example 2.6 *let A, B, and C be sets. We are going to give a very careful justification of all the steps in these examples. Later on, we will become more lax for otherwise mathematical proofs become unmanageably long.*

1. *Prove $A \cap (B \cup C) = (A \cap B) \cup (A \cap C)$.*

   Proof 2.4 *Start with $x \in A \cap (B \cup C)$ and we need to show $x \in (A \cap B) \cup (A \cap C)$. This shows that $A \cap (B \cup C) \subseteq (A \cap B) \cup (A \cap C)$. By definition of intersection, $x \in A \cap (B \cup C)$ means that $x \in A$ and $x \in B \cup C$. By definition of union, $x \in A$ and either $x \in B$ or $x \in C$. By Exercise 4d. in Section 1.1, the previous statement is logically equivalent to "either $x \in A$ and $x \in B$ or $x \in A$ and $x \in C$. But again, by definition of intersection, $x \in A \cap B$ or $x \in A \cap C$, which implies by definition of union that $x \in (A \cap B) \cup (A \cap C)$.*

   To prove $(A \cap B) \cup (A \cap C) \subseteq A \cap (B \cup C)$ one need only read the above argument in reverse. Hence, showing subset both ways implies the equality of the two sets. □

   One comment we wish to make about this proof is that we could have replaced each implication with equivalence or "iff" or we may just write "equivalently", since each step in the proof was justified by either a definition or a logical equivalence. We will do this from now on.

2. *Prove $(A')' = A$.*

   Proof 2.5 *$x \in (A')'$ is equivalent to $x \in U - A'$, which is equivalent to $x \in U$ and $x \notin A'$, which is equivalent to $x \in U$ and $\neg(x \in U$ and $x \notin A)$. By DeMorgan's Law (see Exercise 6 in Section 1.1), this is equivalent to $x \in U$ and either $x \notin U$ or $\neg(x \notin A)$. By Exercise 3a. in Section 1.1, this is equivalent to $x \in U$ and either $x \notin U$ or $x \in A$. By Exercise 4d. in Section 1.1, this is equivalent to either $x \in U$ and $x \notin U$ or $x \in U$ and $x \in A$. Since $x \in U$ and $x \notin U$ is a contradiction (Example 1.4.2 in Section 1.1) and is therefore False, by Exercise 8.2 in Section 1.1, the previous statement is equivalent to $x \in U$ and $x \in A$. But since $x \in U$ is always True, this is equivalent to $x \in A$ (by Exercise 8.1 in Section 1.1) and the proof is demonstrated.* □

3. *Prove $B \subseteq A \cup B$.*

Proof 2.6 *By definition of union we need to show $(x \in B) \rightarrow (x \in A) \vee (x \in B)$, but this is a tautology (see Exercise 2a. in Section 1.1). Thus, the proof is complete.* □

4. *Prove If $A \subseteq B$, then $B = A \cup B$.*

Proof 2.7 *We assume $A \subseteq B$, and we need to show $B = A \cup B$. By the previous example, we already know $B \subseteq A \cup B$, so we need only show that $A \cup B \subseteq B$. Therefore, if $x \in A \cup B$, by definition, $x \in A$ or $x \in B$. Since, by assumption, $x \in A$ implies $x \in B$ is also True, Exercise 5 in Section 1.1, $x \in B$ must also be True, which completes the proof.* □.

5. *Prove $A \subseteq B$ iff $B = A \cup B$.*

Proof 2.8 *Since we are proving an "iff" statement, this requires us to prove implication in both directions. We start by proving $A \subseteq B \rightarrow B = A \cup B$, however, the previous example has already shown this. To prove the reverse implication, we assume $B = A \cup B$ and show that $A \subseteq B$. We do this by element chasing, so assume $x \in A$ is True. By Exercise 2a. in Section 1.1, we know $x \in A$ or $x \in B$ is also True. In other words, $x \in A \cup B$ is True. Now since $B = A \cup B$, by substituting $A \cup B$ with B in the statement "$x \in A \cup B$ is True", we have $x \in B$ is True, which completes the proof.* □

This final step in the last proof just presented brings up another important point when demonstrating a proof which for some may seem self-evident, namely

> **Substitution Principle of Equality:** *If two objects are known to be equal, then we can replace one object with the other in any statement without changing the validity of the statement.*

EXERCISES

1 Rewrite the following set using roster notation:

$$\{\, n \in \mathbb{Z} \mid (0 < n < 10) \,\wedge\, (3|n) \,\wedge\, \exists k \in \mathbb{Z}, \, n = 2k + 1 \,\}$$

2 Rewrite the following set using set builder notation:

$$\left\{ 1, \frac{1}{2}, \frac{1}{3}, \frac{1}{4}, \cdots \right\}$$

3 Give a counterexample for the following statement:

Let A, B, and C be sets. If A and B are disjoint and B and C are disjoint, then A and C are disjoint.

Note: two sets are **disjoint** if their intersection is empty.

4 For any sets A, B, and C, prove the results listed below.

a. $A \cup (B \cap C) = (A \cup B) \cap (A \cup C)$.

b. $A - B = A \cap B'$.

c. $(A \cap B) \cup (A - B) = A$.

d. If A and B are disjoint sets, then $A - B = A$ and $B - A = B$.

e. If $A \subseteq B$, then $A \cap C \subseteq B \cap C$.

f. If $A \subseteq B$, then $A \cup C \subseteq B \cup C$.

g. If $A \subsetneq B$ and $B \subseteq C$, then $A \subsetneq C$.

h. $A \subseteq B$ iff $B' \subseteq A'$.

i. $A \subseteq B$ iff $A \cap B' = \emptyset$.

5 Prove the following statements about symmetric difference:

a. $A \bigtriangleup B = B \bigtriangleup A$.

b. If A and B are disjoint sets, then $A \bigtriangleup B = A \cup B$.

c. $A \cap (B \bigtriangleup C) = (A \cap B) \bigtriangleup (A \cap C)$.

d. $A \bigtriangleup (B \bigtriangleup C) = (A \bigtriangleup B) \bigtriangleup C$.

2.3   CONTRAPOSITIVE PROOF

Contrapositive proof is easy to explain, easy to understand and no giant leap from the direct proof. Indeed, it follows from the logical equivalence

$$P \to Q \equiv \neg Q \to \neg P.$$

In other words,

Proof by Contrapositive:
To prove $P \to Q$ one can equivalently show $\neg Q \to \neg P$.

Such a form of proof is called an **indirect** proof. The other form of indirect proof we shall see in the next section is called **proof by contradiction**.

One might ask why we need proof by contrapositive. It turns out, a proof which may be impossible to demonstrate using a direct proof might be straightforward using proof by contrapositive. A simple example should convince the reader of the power of a contrapositive proof. Consider the following statement: "If the square of an integer is even, then the number itself must also be even". Mathematically stated,

$$\forall n \in \mathbb{Z}, \ n^2 \text{ even} \to n \text{ even}.$$

If we were to attempt to prove this directly, we would start with the assumption that n^2 is even. We could then say, by definition of even, $n^2 = 2k$ for some $k \in \mathbb{Z}$. But then we are kind of stuck: we could say $n = \sqrt{2k}$, but where does that get us? Now watch how smoothly the proof is by contrapositive. The contrapositive statement is "If an integer is **not** even, then the square is **not** even". A simpler way to say that is "If an integer is odd, then it's square is also odd".

We will prove the contrapositive statement, since it is also an implication, using a direct proof. Let's assume that an integer n is odd. By definition of odd, this means $n = 2k + 1$ for some integer k. Now

$$n^2 = (2k + 1)^2 = 4k^2 + 4k + 1 = 2(2k^2 + 2k) + 1.$$

Since $2k^2 + 2k$ is an integer and $n^2 = 2(2k^2 + 2k) + 1$, then by definition of odd, n^2 must be odd. □

Example 2.7 *Let m and n be integers. We prove if $m + n \geq 15$, then either $m \geq 8$ or $n \geq 8$.*

Proof 2.9 *We will prove the contrapositive statement, namely, if $\neg(m \geq 8 \vee n \geq 8)$, then $\neg(m + n \geq 15)$. Equivalently, using DeMorgan's law, we prove if $m < 8$ and $n < 8$, then $m + n < 15$. Assume $m < 8$ and $n < 8$. Since m and n are integers, we can say more, namely, $m \leq 7$ and $n \leq 7$. Therefore, $m + n \leq 14$, which is certainly less than 15.* □

EXERCISES

1 Prove each of the following statements using a contrapositive proof.

   a. For integers m and n, if $5 \nmid (m+n)$, then either $5 \nmid m$ or $5 \nmid n$.

   b. For a given integer n, if $7n^2 + 3n + 4$ is odd, then so is n.

   c. For integers m and n, if $m^2(n^2 - 2n)$ is odd, then m and n must both be odd.

   d. For real numbers x and y, if $x^3 + xy^2 \leq y^3 + x^2y$, then $x \leq y$.

2 Use results you've seen thus far in the text to give a direct proof of the following fact: For positive integers a, b, c, if $a^2 + b^2 = c^2$ and a and b are odd, then c is even.

2.4 PROOF BY CONTRADICTION

The idea behind proof by contradiction can appear to be quite convoluted. The logic is that in order to prove $P \to Q$ is True, one can show it's negation, $\neg(P \to Q)$, is False. In logic, this is sometimes called **the law of the excluded middle** which declares that every statement or its negation must be True. Equivalently, since $\neg(P \to Q)$ is logically equivalent to $P \wedge \neg Q$ (convince yourself of this), one must show $P \wedge \neg Q$ is False. Now, in order to show $P \wedge \neg Q$ is False, we will suppose it were True(!) and show a contradictory statement follows of the form $R \wedge \neg R$, for some statement R. Therefore, we can conclude that $P \wedge \neg Q$ cannot be True and must be False. In summary,

---

Proof by Contradiction:
   To prove $P \to Q$, assume P and $\neg Q$ are True and arrive at a contradictory statement of the form $R \wedge \neg R$.

We shall prove by contradiction that $\sqrt{2}$ is an irrational number. Let's first rephrase the statement as an implication: "If $x = \sqrt{2}$, then $x \notin \mathbb{Q}$". As described above, we begin our proof by supposing to the contrary that $x = \sqrt{2}$ and $x \in \mathbb{Q}$. In other words $x = m/n$ where $m, n \in \mathbb{Z}$ and $n \neq 0$.

We will now state something which is common in mathematical proof, namely

> "without loss of generality", sometimes abbreviated as "WLOG" which basically means that we may make an assumption which does not reduce the generality of our proof. Typically, one needs to justify WLOG.

In our proof, WLOG, we may assume that the fraction m/n is in lowest terms, i.e. m and n have no common factors. Indeed if m/n is not in lowest terms, we can always reduce it so that it is, by cancellation. Now consider $x^2 = m^2/n^2$, i.e. $2 = m^2/n^2$. Equivalently, $m^2 = 2n^2$. By definition of even, we know m^2 is even, which coincidentally (actually, by design) we have shown implies m is even. By definition of even, $m = 2k$ for some integer k. By substitution in $m^2 = 2n^2$, we have $(2k)^2 = 2n^2$ or $4k^2 = 2n^2$. By cancellation, $n^2 = 2k^2$ which by definition implies that n^2 is even, and again by our earlier result implies that n is even. Since m and n are both even, this implies they have a common factor of 2. So on the one hand we assumed m and n had **no** common factors, yet we concluded that m and n **had** a common factor, namely a common factor of 2. This is the contradiction, which implies that our original assumption that $x = \sqrt{2}$ and $x \in \mathbb{Q}$ must be False. Thus, "If $x = \sqrt{2}$, then $x \notin \mathbb{Q}$". must be True, which completes the proof. □

Remark 2.3 *Let's make some remarks and observations about the proof just given regarding proof by contradiction.*

1. *It sometimes is a good idea to start a proof by contradiction with the phrase "Suppose to the contrary that...". Mathematicians sometimes get lazy and simply say "Suppose not", but for the beginner proof writer this can be a bit confusing; it's better to be more formal about the fact that one is demonstrating a proof by contradiction.*

2. *As we pointed out in a proof by contradiction, one arrives at a contradictory statement of the form $R \wedge \neg R$. In the proof above,*

$$R = \text{``} m \text{ and } n \text{ have a common factor''.}$$

3. *Mathematicians sometimes end a proof by contradiction with a special symbol indicating that a contradiction has occurred. Some examples of this are $\Rightarrow\Leftarrow$, $\perp$, or $\frac{}{\iota}$.*

For the next examples of proof by contradiction we need a formal definition of what it means to be a prime number.

Definition 2.7 *A natural number is a* **prime** *if there are exactly two distinct positive numbers which divide it, namely one and itself. More formally, $p \in \mathbb{N}$ is prime if*

$$p > 1 \ \wedge \ \forall n \in \mathbb{N} \ n \mid p \rightarrow (n = 1 \ \vee \ n = p).$$

Example 2.8 *We now give several proofs regarding prime numbers, two of which employ a proof by contradiction. We will present the proofs with the level of detail one would expect for a proof by contradiction.*

1. *We prove that given a positive integer $n > 1$, if $p > 1$ is the smallest positive integer which divides n, then p must be prime.*

   Proof 2.10 *Suppose to the contrary, given a positive integer $n > 1$ with $p > 1$ the smallest positive integer which divides n, that p was* **not** *prime. By definition of prime, p would have at least another positive divisor k different from 1 and p. Appealing to Exercise 5d. in Section 2.1, we know that $1 < k < p$. Since $k \mid p$ and $p \mid n$, by Example 2.2 in Section 2.1, we know that $k \mid n$, but this contradicts the fact that $p > 1$ is the smallest positive integer which divides n. Thus, our proof is complete.* □

2. *We prove that a positive integer p is prime iff there does not exist another prime q which divides p.*

   Proof 2.11 *Set $P =$ "a positive integer p is prime". and $Q =$ "there does not exist another prime q which divides p". To prove the "iff" statement, we need to show $P \rightarrow Q$ and $Q \rightarrow P$. We will prove both of these using a contrapositive proof.*

   First, to prove $P \rightarrow Q$, we prove $\neg Q \rightarrow \neg P$. Assume there is another prime q which divides p. Then by definition of prime, p cannot be prime having at least three divisors 1, q and p, which demonstrates the implication.

   Second, to prove $Q \rightarrow P$, we prove $\neg P \rightarrow \neg Q$. Assume p is not prime. By definition of prime, p has at least one more positive divisor n. Should n be prime, our proof would be complete. In the case that n is not prime, we know n has other positive integer divisors besides 1 and n. Let q be the smallest of all the divisors of n which is greater than 1. By the previous result, we know that q must be prime. Since $q \mid n$ and $n \mid p$, by Example 2.2 in Section 2.1,

$q \mid p$, which demonstrates the implication and completes the proof.
□

We wish to make two comments about this result and its proof. First, this result as an "iff" statement gives us an equivalent way of defining **prime**, and we could have used this as our original definition. Which actually brings up another point which can confound beginning proof writers. It's the fact that a definition is always an "iff" statement, i.e. the definition being defined is always equivalent to the property given in the definition.

Second, the proof just given uses a property of the natural numbers as an assumption called the Well-ordering principle which states that "any non-empty subset of the natural numbers has a smallest element". We used this fact when we found q in the second half of the proof. The non-empty subset of $\mathbb{N}$ is

$$\{m \in \mathbb{N} \mid m > 1 \ \wedge \ m \mid n\}.$$

We will revisit this is a topic later in the text.

3. The following result and its proof first appeared in a work written by the Greek mathematician, Euclid, in a book called Elements (written around 300 BC). It states that there are an infinite number of primes. Let's first rewrite the statement in the form of an implication. One way could be

   "If A is the set of all prime numbers, then $|A| = \infty$".

   Proof 2.12 Suppose to the contrary that A was finite. Then we can express

   $$A = \{p_1, p_2, p_3, \ldots, p_n\}, \ for \ some \ positive \ integer \ n.$$

   Set $q = (p_1 p_2 p_3 \cdots p_n) + 1$.

   Claim 2.1 q is a prime number.

   We prove this claim using a proof by contradiction. Suppose to the contrary that q was not prime. Then by the previous result, there exists a prime which divides q. By assumption, this prime *dividing q is one of the assumed-to-be-complete set of primes* $p_1, p_2, p_3, \ldots, p_n$. *WLOG, we can assume the prime dividing q is* p_n *(by reordering and relabeling the primes – this makes the proof a tad cleaner). This implies $q = p_n k$ for some integer k. By substitution in the definition of q we have*

$$p_n k = (p_1 p_2 p_3 \cdots p_n) + 1 \text{ or equivalently}$$

$$1 = p_n k - (p_1 p_2 p_3 \cdots p_n) = p_n[k - (p_1 p_2 p_3 \cdots p_{n-1})].$$

*By definition of divides, this implies $p_n \mid 1$. However, by Exercise 5d. in Section 2.1, we must conclude that $p_n = 1$, and so p_n is **not** prime, a contradiction. Hence, the claim is proved.*

Returning to the main proof, for $i = 1, 2, 3, \ldots, n$, we have $q = (p_1 p_2 p_3 \cdots p_n) + 1 > p_i$. This implies q is a prime not in A which is assumed to be the complete list of primes, a contradiction. Hence, the proof is complete. □

Some final remarks about this last proof. Oftentimes in a mathematical proof a statement made in the proof may require it's own mathematical proof. Typically, we isolate this statement and call it a "claim" and prove it separately and internally in the larger proof we wish to demonstrate. As it turned out in this particular proof, the claim was also demonstrated by a proof by contradiction!

EXERCISES

1 Prove each of the following statements using proof by contradiction.

   a. $\sqrt[3]{2}$ is an irrational number.

   b. The sum of a rational number and an irrational number is an irrational number.

   c. The product of a non-zero rational number and an irrational number is an irrational number.

2.5   PROOF BY MATHEMATICAL INDUCTION

The method we present here applies to mathematical statements which are a function of the integers beginning at a certain value. We can represent such a statement by $P(n)$, where n is a variable representing an integer. Here is a classic example.

Example 2.9

$$\underbrace{1 + 2 + 3 + \cdots + n = \frac{n(n+1)}{2}}_{P(n)}, \ where \ n = 1, 2, 3, \ldots.$$

This is a statement which describes a formula for finding the sum of the first n positive integers. For example, P(5) is

$$1 + 2 + 3 + 4 + 5 = \frac{5(5+1)}{2}.$$

Check for yourself that the left-hand-side equals the right-hand-side.

The method we will use to prove such statements is called **mathematical induction** and the are two types: **Weak Induction** and **Strong Induction**. As we shall see, the terms *weak* and *strong* have nothing to do with the effectiveness of the induction, but rather refer to how weak or strong of a certain assumption one makes (details to come). Let's begin with Weak Induction.

> **Weak Mathematical Induction:** Consider a statement $P(n)$ and an integer n_0 which represents the starting value for the statement. If you are able to prove
>
> (I) $P(n_0)$ is True, and
>
> (II) For every $n \geq n_0$, $P(n) \rightarrow P(n+1)$ is True, then
>
> $P(n)$ is True for all values of n beginning with the value n_0.

Item I is sometimes called the **base case** or **atomic step**, while item II is called the **inductive step**.

What justifies the fact that proving items I and II allow us to conclude that $P(n)$ is True for all values of n starting at n_0? Before we give a formal mathematical proof justifying the use of Weak Induction, let's first understand intuitively why it's True. Basically, item II says that "whenever $P(n)$ is True for one value, then it's True for the next value", and item I says "$P(n)$ is True for the first value". Using II, it must be therefore True for the second value. Using II, again, it's True for the third. Thus, $P(n)$ must be True for all values starting with the first value. One can visualize induction as dominoes starting at one point and lining up out into infinity. Each domino represents the mathematical statement a particular value. A statement being True at a particular value corresponds to that domino falling. Item I guarantees the first domino will fall while item II implies that all the dominos are close enough to each other so that when one falls it knocks over the next. If this is the case, then we are guaranteed that all the dominos will fall.

We will now prove Example 2.9 by Weak Induction.

Proof 2.13 *First, we verify that item I is True, i.e. $P(1)$ is True. Now $P(1)$ says*

$$1 = \frac{(1)((1)+1)}{2}, \text{ which is certainly True.}$$

Second, we prove the implication $P(n) \to P(n+1)$ is True. We will use a direct proof. Therefore, we assume the statement $P(n)$ is True for some value of $n \geq 1$. In other words, for a particular value of $n \geq 1$,

$$1 + 2 + 3 + \cdots + n = \frac{n(n+1)}{2}.$$

We now need to show that the statement is True for $n+1$, i.e. that $P(n+1)$ is True and this translates as

$$1 + 2 + 3 + \cdots + (n+1) = \frac{(n+1)((n+1)+1)}{2},$$

or equivalently, $1 + 2 + 3 + \cdots + (n+1) = \frac{(n+1)(n+2)}{2}$.

We will prove this equality by beginning on the left-hand-side and arriving at the right-hand-side and making use of the assumption that $P(n)$ is assumed to be True.

$$1+2+3+\cdots+(n+1) = (1+2+3+\cdots+n)+(n+1) = \frac{n(n+1)}{2}+(n+1)$$

$$= \frac{n(n+1))}{2} + \frac{2(n+1)}{2} = \frac{n(n+1)+2(n+1)}{2} = \frac{(n+1)(n+2)}{2}.$$

This verifies that $P(n+1)$ is True, thus item II is True which by Weak Induction show that $P(n)$ is True for any $n \geq 1$. □

Here is an intuitive "proof by picture" of the result just proved. The left-hand-side of $P(n)$ corresponds to the number of squares in Figure 2.2 (a). In part (b) we consider two identical stairs one of which we rotate $180°$ so we can piece the two together. Finally, in part (c) we see the result is an n-by-$(n+1)$ rectangle. Since two stairs has the number of squares equal to $n(n+1)$, a single set of stairs equals half that amount, i.e.

$$1 + 2 + 3 + \cdots + n = \frac{n(n+1)}{2}.$$

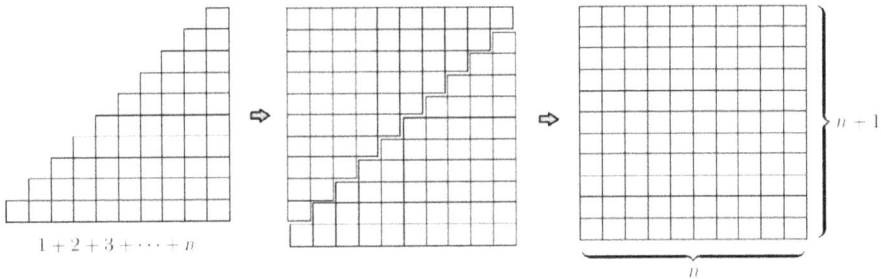

Figure 2.2 (a) The number of squares equals the left-hand-side of $P(n)$. (b) Two stairs put together. (c) The result is an n-by-$(n+1)$ rectangle.

Let's now give the formal proof that Weak Induction does what it claims to do.

Theorem 2.2 *Consider a statement $P(n)$ and an integer n_0. Suppose that*

(I) $P(n_0)$ *is True, and*

(II) *For every $n \geq n_0$, $P(n) \to P(n+1)$ is True.*

Then $P(n)$ is True for all values of n beginning with the value n_0.

Proof 2.14 *The method of proof is by contradiction. Therefore, suppose to the contrary that items I and II are True and the conclusion is False. In other words, there are values of $n \geq n_0$ for which $P(n)$ is False. By the Well-ordering principle, there is a smallest $m \geq n_0$ for which $P(m)$ is False. By item I, $m \neq n_0$, therefore, $m - 1 \geq n_0$ and so $P(m-1)$ is True, since $m - 1$ is smaller than the smallest number m for which the statement is False. But then, by item II, since $P(m-1)$ is True we know that $P((m-1)+1)$ or $P(m)$ is True. However, this contradicts what we said earlier in the proof that $P(m)$ is False.* □

Example 2.10 *We prove that the number of subsets of a set of size n is equal to 2^n for any $n = 0, 1, 2, \ldots$.*

Before we prove this result, let's first introduce notation and terminology for the set of all subsets of a given set.

Definition 2.8 *For a given set A, the* **power set** *of A, written $\mathcal{P}(A)$, is the set of all subsets of A, i.e.*

$$\mathcal{P}(A) = \{B \mid B \subseteq A\}.$$

To restate Example 2.10 mathematically, we wish to show that if A is a set and $|A| = n$, then $|\mathcal{P}(A)| = 2^n$ for any $n \in \mathbb{N}$. First, let's look at a specific instance of this example. If $A = \{a, b, c\}$, then

$$\mathcal{P}(A) = \{\emptyset, \{a\}, \{b\}, \{c\}, \{a, b\}, \{a, c\}, \{b, c\}, A\}.$$

As you can see when $|A| = 3$ we have $|\mathcal{P}(A)| = 8 = 2^3$ as Example 2.10 claims to be True. Now, to prove this result we will make an observation that subsets of A can be divided into two equal and disjoint sets: Ones which contain the element c and ones which do not. Indeed,

$$\mathcal{P}(A) = \{\emptyset, \{a\}, \{b\}, \{a, b\}\} \cup \{\{c\}, \{a, c\}, \{b, c\}, A\}.$$

As you can see, the set $\{\emptyset, \{a\}, \{b\}, \{a, b\}\}$ are the subsets of the set $A - \{c\}$. Furthermore, the set $\{\{c\}, \{a, c\}, \{b, c\}, A\}$ can be obtained by adding the element c to each subset in $\{\emptyset, \{a\}, \{b\}, \{a, b\}\}$.

We are now ready to demonstrate the proof of Example 2.10 using Weak Induction.

Proof 2.15 *For item I, when $n = 0$ this means $A = \emptyset$, and the only subset of the empty set is itself, i.e. $\mathcal{P}(A) = \{\emptyset\}$ and so $|\mathcal{P}(A)| = 1 = 2^0$.*

For item II, we assume the statement is True for all sets of a particular size n and show it is True for a set of size $n + 1$.

Let $A = \{a_1, a_2, \ldots, a_{n+1}\}$. The subsets of A are of two distinct types: Subsets which do not contain a_{n+1} and subsets which do contain a_{n+1}. Now, subsets which do not contain a_{n+1} are precisely the subsets of $\{a_1, a_2, \ldots, a_n\}$ which we know by induction has size 2^n. Furthermore, the subsets of A which do contain a_{n+1} can be obtained by adding a_{n+1} to each subset of $\{a_1, a_2, \ldots, a_n\}$, and therefore also has size 2^n. Hence,

$$|\mathcal{P}(A)| = 2^n + 2^n = 2 \cdot 2^n = 2^1 2^n = 2^{n+1}.$$

□

Example 2.11 *We will prove by Weak induction the following statement: If $x \in \mathbb{R}$ with $x \geq -1$ but $x \neq 0$, then $(1+x)^n > 1 + nx$ for any positive integer $n \geq 2$.*

Proof 2.16 *For item I, we prove the statement is True for $n = 2$, i.e.*

$$(1+x)^2 > 1 + 2x.$$

First, we point out that since $x \geq -1$ but $x \neq 0$, this implies that $x^2 > 0$. Therefore,

$$(1+x)^2 = 1 + 2x + x^2 > 1 + 2x.$$

To verify item II, we assume for a particular $n \geq 2$ that $(1+x)^n > 1 + nx$ and we need to show that $(1+x)^{n+1} > 1 + (n+1)x$. Now,

$$(1+x)^{n+1} = (1+x)(1+x)^n > (1+x)(1+nx)$$

$$= 1 + x + nx + nx^2 > 1 + x + nx = 1 + (n+1)x.$$

□

The next example is an important result called the **Division Algorithm**.

Theorem 2.3 (Division Algorithm) *If m and n are positive integers, then there exist $q, r \in \mathbb{N}$ such that $n = mq + r$ with $0 \leq r < m$.*

Proof 2.17 *This will be a proof by Weak induction using the variable n. First, if $n = 1$, then we are looking for q and r such that $1 = mq + r$ with $0 \leq r < m$. Should $m = 1$, then $q = 1$ and $r = 0$ works. Otherwise, if $m > 1$, then $q = 0$ and $r = 1$ works.*

Now assume the statement is True for a particular integer $n \geq 1$, i.e. there exist $q, r \in \mathbb{Z}$ such that $n = mq + r$ with $0 \leq r < m$. We need to show the statement is True for $n + 1$, i.e. we are looking for q' and r' such that $n + 1 = mq' + r'$ with $0 \leq r' < m$.

Notice that $n + 1 = (mq + r) + 1$. Should $r < m - 1$, then $q' = q$ and $r' = r + 1$ works, since $n + 1 = mq + (r+1)$ with $0 \leq (r+1) < m$. Otherwise, if $r = m - 1$ (the largest value of r), then

$$n + 1 = mq + (r+1) = mq + m = m(q+1).$$

Therefore, $q' = (q+1)$ and $r' = 0$ works. □

Example 2.12 *We present one last example of Weak induction which illustrates it's use when dealing with* **recursively defined sequences,** *also called* **recurrence relations.** *This is a topic you may have seen in a calculus course. Consider the sequence $\{a_n\}$ defined as follows: $a_0 = 1$ and $a_{n+1} = \sqrt{2 + a_n}$ for $n \in \mathbb{N}$. Let generate several terms in the sequence.*

$$a_0 = 1$$

$$a_1 = \sqrt{2 + a_0} = \sqrt{3}$$

$$a_2 = \sqrt{2 + a_1} = \sqrt{2 + \sqrt{3}}$$

$$a_3 = \sqrt{2 + a_2} = \sqrt{2 + \sqrt{2 + \sqrt{3}}}$$

Claim 2.2 $a_n \leq 2$ *for $n \in \mathbb{N}$.*

For the base case, $a_0 = 1 \leq 2$. Now assume that $a_n \leq 2$ for a particular $n \in \mathbb{N}$ and we will show that $a_{n+1} \leq 2$. Notice that

$$a_{n+1} = \sqrt{2 + a_n} \leq \sqrt{2 + 2} = \sqrt{4} = 2.$$

$\square$

Claim 2.3 $\{a_n\}$ *is an increasing sequence, i.e. $a_{n+1} \geq a_n$ for $n \in \mathbb{N}$.*

For the base case, we need to show that $a_1 > a_0$, but this is True since $a_0 = 1$ while $a_1 = \sqrt{3} > 1$. For the inductive step, we assume for a particular $n \in \mathbb{N}$ that $a_{n+1} \geq a_n$ and we will show that $a_{n+2} \geq a_{n+1}$. We shall build up this inequality starting with the one assumed. Refer to Exercise 4 in Section 2.1 for helpful facts. Notice that

$$a_{n+1} \geq a_n \quad \Rightarrow \quad 2 + a_{n+1} \geq 2 + a_n \quad \Rightarrow \quad \sqrt{2 + a_{n+1}} \geq \sqrt{2 + a_n}.$$

But this last equality, by definition, says $a_{n+2} \geq a_{n+1}$. $\square$

There is utility in proving these two claims about a recurrence relation, but the justification is not within the scope of this text. To explain briefly, by the Monotone Convergence Theorem, a sequence satisfying both claims must converge to a value. Indeed, one can show for this example that as n increases, a_n will increase and get closer and closer to the value 2, i.e. in calculus notation

$$\lim_{n \to \infty} a_n = 2.$$

For the inductive step of mathematical induction, sometimes it's not enough to know the previous $P(n)$ implies the next $P(n+1)$. You may need a stronger premise to conclude $P(n+1)$. This type of induction is called **Strong Induction**.

Strong Mathematical Induction: Consider a statement $P(n)$ and an integer n_0 which represents the starting value for the statement. If you are able to prove

(I) $P(n_0)$ is True, and

(II) For every $n \geq n_0$, $[P(n_0) \wedge P(n_0+1) \wedge \cdots \wedge P(n)] \to P(n+1)$ is True, then

$P(n)$ is True for all values of n beginning with the value n_0.

As it turns out, making this stronger premise in the inductive step does not alter the validity of the conclusion that $P(n)$ is True for all $n \geq n_0$. We leave the proof (nearly identical to the Weak Induction proof) of this fact as an exercise. Let's now look at some example applications of Strong Induction. The first result is an important fact about prime numbers.

Theorem 2.4 (The Fundamental Theorem of Arithmetic) *Every natural number larger than 1 is either prime or can be represented as a product of primes.*

Proof 2.18 *Let's be more formal about the statement of the theorem which says if $n \geq 2$, then $n = p_1 p_2 \cdots p_m$ for some positive integer m and prime numbers $p_1, p_2, \ldots, p_m$.*

For the base case $n = 2$, the number 2 is itself the prime p_1 with $m = 1$.

For the inductive step, we assume any natural number larger than 1 and less than or equal to n can be represented as a product of prime numbers, and we seek to show that $n+1$ can be represented as a product of prime numbers. Now should $n+1$ be prime, then as in the base case, $n+1$ is p_1 and $m = 1$. Otherwise, $n+1$ is not prime, in which case $n+1$ has a divisor strictly between 1 and $n+1$, i.e. $n+1 = ab$ for some positive integers a and b with $1 < a < n+1$. Note that b must also be strictly between 1 and $n+1$ since it, too, divides $n+1$. Since both a and b are larger than 1 and less than or equal to n, each is either prime or *can be represented as a product of prime numbers, i.e.*

$$a = p_1 p_2 \cdots p_m \quad and \quad b = q_1 q_2 \cdots q_r,$$

where the p_i's and the q_i's are prime numbers and m and r are positive integers. But then

$$n + 1 = p_1 p_2 \cdots p_m q_1 q_2 \cdots q_r,$$

and so $n+1$ has been represented as a product of prime numbers. Since we have verified items I and II in the statement of Strong Induction, the statement of the theorem must be True. □

Example 2.13 *The following example shows that any positive integer can be represented as a binary number. Formally, we show that for any integer $n \geq 1$, there exists a $k \in \mathbb{N}$ such that*

$$n = a_0 2^0 + a_1 2^1 + a_2 2^2 + \cdots + a_k 2^k, \quad where \ a_0, a_1, a_2, \ldots, a_k \in \{0, 1\}.$$

Indeed, the binary representation of n is $a_0 a_1 a_2 \cdots a_k$.

Proof 2.19 *For the base case, if $n = 1$, then $a_0 = 1$ and $k = 0$.*

For the inductive step, assume all integers from 1 up to n can be represented as a binary number, and we wish to show the same is True about $n+1$. Using Theorem 2.3 with $m = 2$, we have $n+1 = 2q+r$ where $0 \leq r < 2$, i.e. $r \in \{0, 1\}$. Certainly, $q < n+1$, therefore by assumption, $q = a_0 + a_1 2 + a_2 2^2 + \cdots + a_k 2^k$ where $a_0, a_1, a_2, \ldots, a_k \in \{0, 1\}$. But then

$$n + 1 = 2q + r = 2(a_0 2^0 + a_1 2^1 + a_2 2^2 + \cdots + a_k 2^k) + r$$

$$= r 2^0 + a_0 2^1 + a_1 2^2 + a_2 2^3 + \cdots + a_k 2^{k+1},$$

but this last expression is exactly the binary representation of $n + 1$. □

Note that there is nothing special about base 2, for the very same proof can be used for any integer base larger than 2 as well.

Example 2.14 *Consider the sequence $\{a_n\}$ defined as follows: $a_0 = 1$, $a_1 = 2$ and $a_{n+2} = \frac{1}{2}(a_n + a_{n+1})$ for $n \in \mathbb{N}$. We will show that every element of the sequence is between 1 and 2, i.e. $1 \leq a_n \leq 2$ for $n \in \mathbb{N}$.*

Proof 2.20 There are two base cases here, since each element of the sequence is generated by the previous two. However, certainly $1 \leq a_0$, $a_1 \leq 2$.

For the inductive step we assume all elements of the sequence from a_0 up to a_{n+1} are between 1 and 2, and we will show that the same is True for a_{n+2}. By assumption $1 \leq a_n \leq 2$ and $1 \leq a_{n+1} \leq 2$. By adding these two inequalities we have

$$2 \leq a_n + a_{n+1} \leq 4.$$

Multiplying through by $\frac{1}{2}$ yields $1 \leq \frac{1}{2}(a_n + a_{n+1}) \leq 2$ which by how this sequence is defined says $1 \leq a_{n+2} \leq 2$. □

2.5.1 Misuses of Mathematical Induction

This subsection is more for fun than anything else, and can be skipped without losing any continuity, however it may add additional insight into how induction works. We will be attempting to prove some outrageous statements using mathematical induction. Will the reader be able to find the flaw in the logic? We will not specifically identify the flaw and leave the matter as a point of discussion.

Example 2.15 *These first examples will employ Weak Induction.*

1. *We will show that all days are sunny days.*

   Proof 2.21 The base case is certainly True, for there is certainly one day which is sunny. For the inductive step we will assume any sequence of n consecutive days is sunny and show the same is True for $n + 1$ days. To do this, notice that $n + 1$ consecutive days are made up of two overlapping sequence of n consecutive days. Indeed, if we label the $n + 1$ consecutive days as $d_1, d_2, \ldots, d_n, d_{n+1}$, then $d_1, d_2, \ldots, d_n$ and $d_2, d_2, \ldots, d_{n+1}$ are each n consecutive days which by assumption are all sunny days. Therefore, $d_1, d_2, \ldots, d_n, d_{n+1}$ are also all sunny days. □

2. *We will show that everyone is bald.*

   Proof 2.22 Let's be more formal about what we wish to prove. We shall show that the head of a person with n hairs is bald, where $n \in \mathbb{N}$.

For the base case, $n = 0$, certainly a person with no hair is bald. For the inductive step, assume a person with n hairs is bald and we will show the same is True for $n + 1$ hairs. However, this is certainly True, because adding one additional hair on a person's head can by no means change the state of "baldness" to one of non-"baldness". □

3. *A man charged with a heinous crime was brought before a king. This crime was so heinous that a judgement of beheading was made. In addition, since the crime was so completely heinous, the king decided that judgement would be passed within the next 30 days, however that day would not be known to the criminal so that the execution would come as a complete surprise to the criminal. Indeed, each day the man would be brought out to the guillotine with his neck put in place. If it was not the predetermined day, then he would be returned to his cell. This procedure would be repeated each day until the ill-fated day when he would actually lose his head. Using mathematical induction the criminal proved that he would not be beheaded.*

   Proof 2.23 *The criminal proved the following statement: "The execution would not occur on the last n days" where n is an integer between 1 and 30.*

   For the base case, $n = 1$, the criminal knew he would not be executed on the very last day, since than it would not be a surprise. Hence $n = 1$ is True.

   For the inductive step, assuming the execution would not occur on the last n days, he proceeded to show the execution would not occur on the last $n + 1$ days. He reasoned, since the execution would not occur on the last n days, it could not occur at the day just prior to those last n days, for then it would not be a surprise. □

Example 2.16 *These examples will employ Strong Induction.*

1. *We will show for any real number $a \neq 0$ that $a^n = 1$ for every $n \in \mathbb{N}$.*

Proof 2.24 *For the base case, certainly* $a^0 = 1$. *Assume now the statement is True for all natural numbers up through* n. *We will show the statement is True for* $n + 1$.

$$a^{n+1} = \frac{a^n a^n}{a^{n-1}} = \frac{1 \cdot 1}{1} = 1.$$

□

2. *We will show all natural numbers are interesting.*

Proof 2.25 *For the base case, no one can argue that zero is an interesting number. For instance, it is a number which represents nothing. It's also the only natural number which every integer divides. The list could go on as to why zero is so very interesting.*

For the inductive step, assume all the natural numbers up through n *are interesting and we will show that* $n+1$ *is also interesting. We will prove this by contradiction. Suppose* $n+1$ *was not interesting. Then* $n + 1$ *would be the smallest uninteresting natural number. But that's interesting! So assuming* $n + 1$ *is not interesting leads to a contradiction. Therefore,* $n + 1$ *must be interesting.* □

EXERCISES

1 Prove each of the following statements using proof by Weak induction.

a. Show

$$1^2 + 2^2 + 3^2 + \cdots + n^2 = \frac{n(n + 1)(2n + 1)}{6}.$$

b. Show

$$1^3 + 2^3 + 3^3 + \cdots + n^3 = \frac{n^2(n + 1)^2}{4}.$$

c. Show that $3 \mid (4^n - 1)$ for $n \in \mathbb{N}$.

d. Show that $5 \mid (11^n - 6)$ for any positive integer n.

e. Show that $6 \mid (n^3 - n)$ for any positive integer n.

f. If A is a set with at least two elements, then the number of subsets of A of size 2 equals $\frac{n(n-1)}{2}$ where $|A| = n$.

    g. For $n \geq 4$ an integer, $n! > 2^n$.

    h. For $n \geq 4$ an integer, $2^n \geq n^2$.

2 Explain why

$$1^3 + 2^3 + 3^3 + \cdots + n^3 = (1 + 2 + 3 + \cdots + n)^2.$$

3 Prove using Weak induction and an earlier exercise that, for sets $A_1, A_2, \ldots, A_n$,

$$(A_1 \cup A_2 \cup \cdots \cup A_{n-1}) \cap A_n = (A_1 \cap A_n) \cup (A_2 \cap A_n) \cup \cdots \cup (A_{n-1} \cap A_n).$$

4 Provide a proof, similar to the one for Weak induction, to justify the validity of Strong Induction.

5 Prove each of the following statements using proof by Strong induction.

    a. Show any integer greater than 1 is divisible by a prime number (without using Theorem 2.4).

    b. Consider the following recurrence relation:

$$a_0 = 0, \ a_1 = 1, \ \text{and} \ a_{n+2} = a_n + a_{n+1} \ \text{for} \ n \in \mathbb{N}.$$

    This sequence is called the **Fibonacci** sequence after the Italian mathematician Leonardo of Pisa, also known by the alias Fibonacci. It's a curious thing that some mathematicians took on pseudonyms. There are lots of wonderful mathematical facts about the Fibonacci sequence. For this exercise show that $a_n < 2^n$ for all $n \in \mathbb{N}$.

2.6 PROOF BY CASES

There are times when it is not possible to prove a mathematical statement all in one shot and in all its generality. In this situation it is sometimes useful to break up the proof into several cases. These cases may be completely non-intersecting, and at other times later cases may rely on the verification of earlier cases. The formal situation is sometimes called

a **divide and conquer** proof by cases, while the later is sometimes
called a **bootstrap** proof by cases.

We point out that when demonstrating a divide and conquer proof,
make sure that all the individual cases together cover all possible cases.
For the divide and conquer proof by cases, we will prove some facts about
absolute value. Let's give the formal definition.

Definition 2.9 *For a real number a, the* **absolute value** *of a, written*
$|a|$*, is defined in cases.*

$$|a| = \begin{cases} a, & \text{if } a \geq 0 \\ -a, & \text{if } a < 0 \end{cases}$$

Note that such a definition is called a **branch** *function or a* **piecewise defined** *function. We will talk a bit more about these functions
later on in the text.*

This brings up a good point. It may not always be clear how to
divide into cases, however there are sometime clues as to how to do
it. For instance, the following facts about absolute value may naturally
divide into cases according to the two cases in the definition of absolute
value. In practice, you may need to adjust your cases while crafting your
proof.

Example 2.17 *Let a and b be real numbers,*

1. $|a| \geq 0$.

2. $a \leq |a|$.

3. $|a| = |-a|$.

4. $|ab| = |a||b|$.

Proof 2.26 *For the first statement, we will need two cases. The cases
follow exactly the branches of the definition of absolute value.*
   Case 1: $a \geq 0$
   In this case, by the definition of absolute value, $|a| = a$. *Now, because
of the case we're in,* $|a| = a \geq 0$, *and the statement has been proved in
this case.*

Case 2: $a < 0$

In this case, by the definition of absolute value, $|a| = -a$. Now, because of the case we're in, namely $a < 0$, we know that $-a > 0$. Therefore,

$$|a| = -a > 0 \geq 0.$$

Since a real number is either non-negative or negative, our divide and conquer proof is complete. □

The second statement we leave as an exercise.

For the third statement we will need three cases.

Case 1: $a > 0$

In this case, $-a < 0$ and using the definition of absolute value,

$$|-a| = -(-a) = a = |a|.$$

Case 2: $a = 0$

In this case, $-0 = 0$, so certainly $|-0| = |0|$.

Case 3: $a < 0$

In this case, $-a > 0$ and using the definition of absolute value,

$$|-a| = -a = |a|.$$

Since a real number is either positive, zero, or negative, our divide and conquer proof is complete. □

The fourth statement will need more cases, since there are two real numbers involved. Since x and y are arbitrary and $xy = yx$ for any real numbers x and y, WLOG, we will be able to reduce the number of cases. This proof also illustrates how one sometimes verifies an equality, by showing the left-hand-side of the equality and the right-hand-side of the equality equal the same thing, as opposed to starting on one side of the equality and arriving at the other side through a series of equalities.

Case 1: *At least one of a and b equal zero.*

WLOG, we may assume that $a = 0$. Then $ab = 0$ as well and

$$|ab| = |0| = 0 \quad \text{while} \quad |a||b| = |0||b| = 0 \cdot |b| = 0.$$

Since $|ab| = 0$ and $|a||b| = 0$, they must be equal to each other, i.e. $|ab| = |a||b|$.

Case 2: *Both a and b are positive.*
In this case ab is also positive, and so

$$|ab| = ab = |a||b|.$$

Case 3: *Both a and b are negative.*
In this case ab is positive, and so

$$|ab| = ab \quad while \quad |a||b| = (-a)(-b) = ab.$$

Since $|ab| = ab$ and $|a||b| = ab$, they must be equal to each other, i.e.
$|ab| = |a||b|$.
Case 4: *a and b have opposite signs.*
WLOG, we may assume $a > 0$ and $b < 0$. Then $ab < 0$ and

$$|ab| = -(ab) = -ab \quad while \quad |a||b| = a(-b) = -ab.$$

Since $|ab| = -ab$ and $|a||b| = -ab$, they must be equal to each other,
i.e. $|ab| = |a||b|$.
These four cases cover all the possible cases for a and b and therefore
the proof is complete. ☐

Example 2.18 *We will prove that if n is an even number, than n has*
one of the following three forms: $6k$, $6k + 2$, or $6k + 4$, for some integer
k.

Proof 2.27 *Since n is even, we know $n = 2m$, for some integer m. By*
the division algorithm, $m = 3k + r$ where $0 \leq r < 3$. In other words,
$r = 0, 1$, or 2. Therefore, $m = 3k$, $3k + 1$, or $3k + 2$. But then

$$n = 2(3k) = 6k, \quad n = 2(3k + 1) = 6k + 2, \quad or \quad n = 2(3k + 2) = 6k + 4.$$

☐

Example 2.19 *These examples illustrate the bootstrap cases proof*
method.

1. *Consider a real valued function with the additive property*

$$f(x + y) = f(x) + f(y).$$

We will show that $f(rx) = rf(x)$ for all $r \in \mathbb{Q}$.

Proof 2.28 *In this bootstrap proof, we will build up slowly to $r \in \mathbb{Q}$ through a series of cases beginning with the case when r is a natural number.*

Case 1: $f(nx) = nf(x)$ *for all $n \in \mathbb{N}$.*

We will prove this by induction. For the base case, we need to show that $f(0) = 0$, but this is True since

$$f(0) = f(0+0) = f(0) + f(0),$$

and subtracting $f(0)$ from both sides of this equation yields $0 = f(0)$, which proves the base case. For the inductive step, assuming $f(nx) = nf(x)$ for a particular natural number n, then

$$f((n+1)x) = f(nx+x) = f(nx) + f(x) = nf(x) + f(x) = (n+1)f(x).$$

Thus, Case 1 has be proved. Now we extend our proof to integers.

Case 2: $f(nx) = nf(x)$ *for all $n \in \mathbb{Z}$.*

By Case 1, it's enough to show Case 2 is True for $n < 0$. Notice that, since $-n > 0$,

$$0 = f(0) = f(nx+(-n)x) = f(nx) + f((-n)x) = f(nx) + (-n)f(x).$$

For the last step, we appealed again to Case 1. Now add $nf(x)$ to both sides of this equation to get $nf(x) = f(nx)$, which completes the proof of this case. In the next case we consider reciprocals of non-zero integers.

Case 3: $f\left(\frac{1}{n}x\right) = \frac{1}{n}f(x)$ *for all non-zero $n \in \mathbb{Z}$.*

Notice that for any non-zero integer n, using Case 2,

$$f(x) = f\left(n \cdot \frac{1}{n}x\right) = nf\left(\frac{1}{n}x\right).$$

Multiplying both sides of this equation by $\frac{1}{n}$, we have $\frac{1}{n}f(x) = f\left(\frac{1}{n}x\right)$, which proves this case. Now, finally we can prove the complete result for any rational number.

Case 4: $f(rx) = rf(x)$ *for all* $r \in \mathbb{Q}$.

For $r \in \mathbb{Q}$, *express* $r = \frac{m}{n}$ *where* $m, n \in \mathbb{Z}$ *and* $n \neq 0$. *Using Cases 2 & 3,*

$$f(rx) = f\left(\frac{m}{n}x\right) = f\left(m \cdot \frac{1}{n}x\right) = mf\left(\frac{1}{n}x\right)$$

$$= m \cdot \frac{1}{n}f(x) = \frac{m}{n}f(x) = rf(x).$$

□

2. *We will show that* $\cos x \leq \frac{\sin x}{x} \leq 1$ *for any non-zero* x *between* $-\pi/2$ *and* $\pi/2$. *This result is used to prove a well known limit in calculus, namely*

$$\lim_{x \to 0} \frac{\sin x}{x} = 1.$$

Proof 2.29 *We will first prove it for positive values of* x *and then use that case to prove the case of negative values of* x.

Case 1: $0 < x < \pi/2$

Consider Figure 2.3 consisting of several triangles and the unit circle. You will need a bit of geometry and trigonometry to prove this result.

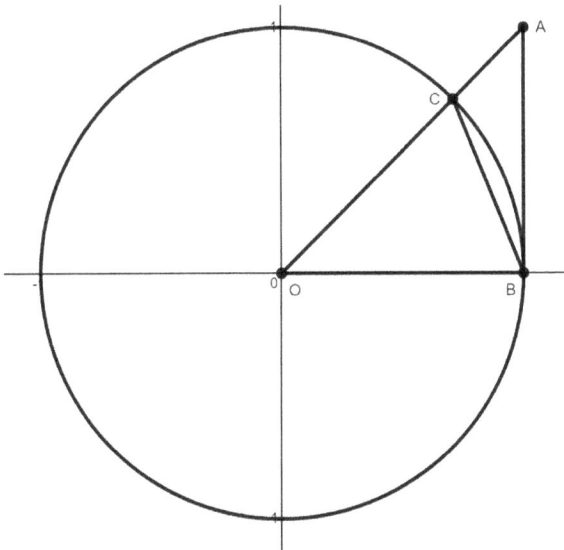

Figure 2.3 The unit circle with several triangles

Suppose $\angle COB$ has value x (in radians). Then the coordinates of C are $(\cos x, \sin x)$ and the height of $\triangle COB$ is $\sin x$ and the height of $\triangle AOB$ is $\tan x$. Now the area of $\triangle COB$ is

$$\frac{1}{2}(1)(\sin x) = \frac{\sin x}{2},$$

and the area of $\triangle AOB$ is

$$\frac{1}{2}(1)(\tan x) = \frac{\tan x}{2}.$$

The area of sector COB is

$$\frac{1}{2}(1)^2(x) = \frac{x}{2}.$$

Now since the area of $\triangle COB$ is less than the area of sector COB which in turn is less than the area of $\triangle AOB$, we have

$$\frac{\sin x}{2} \leq \frac{x}{2} \leq \frac{\tan x}{2}.$$

Multiplying this inequality through by $\frac{2}{\sin x}$ ($\sin x \neq 0$ in this case) gives

$$1 \leq \frac{x}{\sin x} \leq \frac{1}{\cos x}.$$

Taking reciprocals reverses the inequality (see Exercise 4d. in Section 2.1) and gives us the result.

Case 3: $-\pi/2 < x < 0$

Then $0 < -x \leq \pi/2$, so using Case 2 which has already been proved,

$$\cos(-x) \leq \frac{\sin(-x)}{(-x)} \leq 1.$$

We know from trigonometry that $\cos(-x) = \cos x$ and $\sin(-x) = -\sin x$, so making those substitutions into our inequality and canceling negative numbers we arrive at the result. □

3. *Consider a circle with center O, and let A, B, and C be any three distinct points on the circle. Set $\alpha = \angle ABC$ and $\beta = \angle AOC$. Show that $\beta = 2\alpha$. We will leave this as an exercise, but we will give you the case break down.*

   Case 1: *The point O lies on the line segment $\overline{AB}$.*

   Case 2: *The point O is within angle α. (you will need Case 1)*

   Case 3: *The point O is outside of angle α. (you will need both Case 1 & 2)*

EXERCISES

1 Prove each of the following statements using a divide and conquer proof by cases.

   a. Prove the second statement in Example 2.17.

   b. Prove if $a \in \mathbb{R}$, then $a + |a| \geq 0$.

   c. For $a \in \mathbb{R}$ and $b \in \mathbb{R} \geq 0$, if $|a| \leq b$ then $-b \leq a \leq b$. (use cases $a \geq 0$ and $a < 0$).

   d. Prove the Triangle Inequality, namely

   $$\forall x, y \in \mathbb{R}, \ |x + y| \leq |x| + |y|.$$

   e. If n is even, then either $n = 4k$ or $n = 4k + 2$ for some $k \in \mathbb{Z}$.

   f. If n is odd, then $n = 6k + 1$, $n = 6k + 3$ or $n = 6k + 5$ for some $k \in \mathbb{Z}$.

2 Prove each of the following statements using a bootstrap proof by cases.

   a. Let f be a real-valued function satisfying the following properties:
   Property 1. $f(0) \neq 0$, and
   Property 2. $\forall x, y \in \mathbb{R} \quad f(x + y) = f(x)f(y)$.

   Prove that $\forall x \in \mathbb{R}, \forall r \in \mathbb{Q} \quad f(rx) = f(x)^r$.

   b. Prove Example 2.19.3.

2.7   REVIEW OF PROOF METHODS

In this section, we will quickly summarize all the proof methods we have covered in this chapter and present some rules of thumb on when to use them, after which you will be presented with a series of exercises each of which will combine several of the methods you have learned.

When mathematically proving a result:

1. Express the statement as an implication.

2. Try a direct proof first.

3. Next, consider the contrapositive.

4. A proof by contradiction should be your last resort.

5. If you are struggling to demonstrate the proof in all its generality, perhaps splitting the proof into cases may help.

6. If the statement is expressed in terms of integers, try a proof by induction.

7. If proving a result by induction, it is not necessary to decide in advance whether you need weak or strong induction. When you begin proving the inductive step this will become apparent.

We remark that there is a tendency for beginning students to want to apply proof by contradiction to every statement they need to prove. We recommend that you avoid this habit and first consider a more direct approach. One reason is for aesthetic reasons, since the proof by contradiction is considered by many to be less elegant and more convoluted. Indeed, if you can prove a result directly it is easier to follow. In addition, as we explained, the validity of proof by contradiction assumes the **law of the excluded middle**, that every statement is either True or False. Should you explore set theory and mathematical logic in more depth, you would find that this is not always the case. Indeed, it has been shown that no matter what axiom system you assume for the theory of mathematics, there will always be statements that arise that can neither be proved True nor False. In fact, there are some mathematicians/logicians, called intuitionists, who refuse to use proof by contradiction and consider it invalid except in only certain instances.

When tackling the exercises below, we recommend coming up with a game plan before diving in. In other words, sketch out an outline for how you plan to attack each problem. For the exercises which have parts, the earlier parts are meant to assist in proving later parts.

EXERCISES

1 Prove the following statements:

    a. For any $n \in \mathbb{Z}$, if $3|n^2$ then $3|n$

    b. $\sqrt{3}$ is an irrational number.

2 Prove that for $m, n \in \mathbb{Z}$, if $8 \nmid (m^2 - n^2)$, then either m is even or n is even.

3 Let p be a prime number and $m \in \mathbb{N}$

    a. Prove that for all $n = 2, 3, \ldots$, if $p \mid m^n$ then $p \mid m$.

    b. Prove that $\sqrt[n]{p}$ is irrational, for any $n = 2, 3, \ldots$.

4 Prove the following statements:

    a. For integers m and n, if $m^2(n^2 - 2n)$ is odd, then m and n must be odd.

    b. Let n be a positive integer. If the remainder when dividing n by 3 we get a remainder of 2, then n is not the square of another positive integer.

CHAPTER 3

Special Proof Types

I N THIS CHAPTER we focus on special cases of proofs. This includes looking at constructive versus non-constructive proofs, existence proofs and uniqueness proofs. Constructive proofs are important
to identify for they often can lead to algorithmic methods for constructing mathematical objects of interest which make up, for instance, the
topics of computational algebra and computational geometry. Proofs of
existence and uniqueness are also important to focus on, since they tend
to follow a similar pattern of proof and thus covering this topic can in-
struct the aspiring mathematician on a general approach to attacking
them.

3.1 EXISTENCE PROOFS

The aim of an existence proof is to demonstrate a statement of the form
$\exists x \, P(x)$. There are two general ways one goes about proving an existence
statement. One method of proving an existence statement is **constructive**; by this we mean either you construct a particular x which satisfies
$P(x)$, or you derive a general recipe, i.e. an algorithm, for producing x's
which satisfy $P(x)$. In this section we will illustrate both approaches.
Computational mathematics capitalizes on these constructive proofs to
create computer programs which can derive the x in an existence statement. Computational mathematics is a burgeoning field, due in part to
the fact that computers are becoming faster and faster at doing computation. The other method for proving an existence statement is **non-
constructive**; by this we verify the existence of an x satisfying $P(x)$
without explicitly deriving the x. In other words, the proof does not
shed light on what exactly that x might be or how to generate it.

DOI: 10.1201/9781032687728-3

3.1.1 Non-Constructive Proofs

We will begin with existence proofs which are non-constructive. We will now prove a more general version of the Division Algorithm. The first version of this theorem was only for positive integers and the method of proof was by Weak Induction. This more general version will extend to integers.

Theorem 3.1 (Division Algorithm) *If m, n are integers with $m \geq 1$, then there exist $q, r \in \mathbb{Z}$ such that $n = mq + r$ with $0 \leq r < m$.*

Proof 3.1 *Consider the following set:*

$$S = \{n - mx \ : \ x \in \mathbb{Z} \ \text{and} \ n - mx \geq 0\}.$$

First, note that S is a non-empty set, since

$$n - m(-|n|) = n + m|n| \geq n + |n| \geq 0.$$

For the last inequality above, see Exercise 1b. in Section 2.6. Second, by definition, S is a subset of the natural numbers. Therefore, by the Well-ordering principle S has a smallest element. Let $r \geq 0$ be that smallest element in S. Since $r \in S$ we know $r = n - mq$ for some $q \in \mathbb{Z}$. Thus, $n = mq + r$ with $r \geq 0$.

It remains to show that $r < m$. To see this, notice that

$$r - m = (n - mq) - m = n - m(q + 1).$$

So $r - m$ has the proper form to be an element of S. However, $r - m < r$, and since r is the smallest element of S, it must be the case that $r - m < 0$, i.e. $r < m$. □

Example 3.1 *Let's look at a specific case of the proof just given. Set $n = 7$ and $m = 3$. Then*

$$S = \{7 - 3x \ : \ x \in \mathbb{Z} \ \text{and} \ 7 - 3x \geq 0\}.$$

*By plugging in values we see that $S = \{1, 4, 7, \ldots\}$ and that $r = 1$ so that $7 = 3x + 1$ with $0 \leq r < 3$. From this we see that $q = 2$. The reader might then argue that the proof is in fact constructive, but in reality it is not, since there is no effective way of algorithmically finding the smallest element r in S. Without going into too much detail, this falls under a topic in mathematics called **computability** which takes us too far afield for this introductory text.*

For the next non-constructive proof we first need to define some terminology.

Definition 3.1 *Let a and b be two positive integers. The integer d is the* **greatest common divisor** *of a and b, written d = gcd(a, b) if*

1. $d > 0$

2. $d|a$ and $d|b$ (common divisor) and

3. $e|a$ and $e|b$ implies $e|d$ (greatest).

Example 3.2 $\gcd(120, 36) = 12$.

Before we can give an existence proof of the greatest common divisor, we need some preliminary terminology and results. The integer multiples of $n \in \mathbb{Z}$ will be denoted by $n\mathbb{Z}$. For instance,

$$3\mathbb{Z} = \{0, \pm 3, \pm 6, \pm 9, \ldots\}.$$

In the next lemma we will prove some useful facts about $n\mathbb{Z}$ which will be applied to prove the existence of the greatest common divisor.

Lemma 3.1 *Consider the set $n\mathbb{Z}$ for some $n \in \mathbb{N}$. If X is any non-empty subset of the integers closed under addition and subtraction, i.e. $a, b \in X$ implies $a + b, a - b \in X$, then $X = n\mathbb{Z}$ for some $n \in \mathbb{N}$.*

Proof 3.2 *Should $X = \{0\}$, then $X = 0\mathbb{Z}$ and we are done. Otherwise X has positive elements. Indeed, if we take any non-zero element $m \in X$. Then $0 = m - m \in X$ by assumption, and so $-m = 0 - m \in X$ again by assumption. Since m and $-m$ are in X and one of these must be positive we can conclude X has positive elements. Now let n be the smallest positive element in X. We show that $X = n\mathbb{Z}$. First take $x \in n\mathbb{Z}$ so that $x = nk$. If $k = 0$, then $x = 0$ and as we saw above $0 \in X$. If k is positive, then $x = \underbrace{n + \cdots + n}_{k} \in X$ by assumption (and induction). If k is negative, then as we saw above $-n \in X$ and $x = \underbrace{(-n) + \cdots + (-n)}_{-k} \in X$ by assumption and induction. Hence $n\mathbb{Z} \subseteq X$. For the reverse inclusion, take any $x \in X$ and write $x = nq + r$ with $0 \le r < n$ using the Division Algorithm. Since $n\mathbb{Z} \subseteq X$ and X is closed under addition and subtraction, it follows that $r = x - nq \in X$. Since n is the smallest positive integer in X, it must be that $r = 0$ and so $x = nq \in n\mathbb{Z}$.*

Theorem 3.2 *For any non-zero integers a and b, the greatest common divisor of a and b exists.*

Proof 3.3 *Consider the set* $C = \{ax + by \ : \ x, y \in \mathbb{Z} \}$. *Certainly* $C \neq \emptyset$ *(since for instance* $a = a \cdot 1 + b \cdot 0 \in C$*). Furthermore, C is closed under addition and subtraction, since*

$$(ax_1 + by_1) \pm (ax_2 + by_2) = a(x_1 \pm x_2) + b(y_1 \pm y_2) \in C.$$

Therefore, by Lemma 3.1, we know that $C = d\mathbb{Z}$ *for some positive integer d. We show now that* $d = \gcd(a, b)$*. We already have* $d > 0$*. Since* $a, b \in C = d\mathbb{Z}$ *we can write* $a = dk$ *and* $b = dl$ *for some integers k and l which implies* $d|a$ *and* $d|b$*. Finally, suppose* $e|a$ *and* $e|b$*. Since* $d = d \cdot 1 \in d\mathbb{Z} = C$ *we can write* $d = ax_0 + by_0$ *for some* $x_0, y_0 \in \mathbb{Z}$*. Hence, by Lemma 4.3,* $e|(ax_0 + by_0)$*, i.e.* $e|d$*.* □

3.1.2 Constructive Proofs

As we mentioned at the start of this section, a constructive proof gives an explicit way of deriving the x in an existence statement $\exists x \ P(x)$. In this subsection we will give several important illustrations of this type of proof. For the first result, we need to define some terminology.

Definition 3.2 *Given a positive integer n, the* **factorial** *of n or n* **factorial***, written*
$$n! = (1)(2)(3) \cdots (n).$$

In other words, n factorial is the product of the consecutive positive integers up to and including n. So, for instance $5! = (1)(2)(3)(4)(5) = 120$. By convention (a phrase mathematicians use), we define $0! = 1$.

Remark 3.1 *It's a curious thing this notion of defining mathematical concepts "by convention", for oftentimes these conventions lack intuition. Take the case of* $0!$ *which appears incongruent to the way any positive integer's factorial is computed. Another simple example which comes to mind is exponentiation. For real number a and positive integer n,*

$$a^n = \underbrace{a \cdot a \cdots a}_{n \ times},$$

whereas $a^0 = 1$ *which makes no intuitive sense. In [8], the authors define "by convention" as follows: "mathematical truths are true by convention,* *in an attenuated sense that's consistent with realism". We will take this to mean that defining these special instances of a definition "by convention" makes the overall definition consistent with mathematical results. Take the case of exponentiation. One mathematical result is that $a^m a^n = a^{m+n}$ for any integers m and n. In the special case that m or n is zero, the only way to make this result mathematically consistent is to define $a^0 = 1$. Similar mathematical results concerning factorial require $0! = 1$. We will not discuss these in detail, but point the reader to the combinatorial operation of counting the number of ways of choosing k objects from n distinct objects, denoted by $C(n, k)$. The formula for $C(n, k)$ requires for consistency that $0! = 1$.*

Theorem 3.3 *For any positive integer n, there exists n consecutive positive integers for which none of them are prime numbers.*

Proof 3.4 *For a given positive integer n, consider the following n consecutive positive integers:*

$$(n + 1)! + 2, (n + 1)! + 3, (n + 1)! + 4, \ldots, (n + 1)! + (n + 1).$$

We now show they are all not prime numbers. Each has the form $(n + 1)! + k$ where $2 \leq k \leq (n + 1)$ and

$$(n + 1)! + k = k[(1)(2) \cdots (k - 1)(k + 1) \cdots (n + 1) + 1],$$

Since $(1)(2) \cdots (k - 1)(k + 1) \cdots (n + 1) + 1$ is an integer, we see that k divides $(n + 1)! + k$, and therefore $(n + 1)! + k$ cannot be prime. □

As the theorem attests, although (as we've seen) there are an infinite number of prime numbers, there are gaps between consecutive prime numbers as big as we wish!

Example 3.3 *Let's illustrate Theorem 3.3 with a constructive example. Let's take the case of $n = 4$. Then an example of four consecutive non-primes are*

$$5! + 2, 5! + 3, 5! + 4, \ldots, 5! + 5 \quad or \quad 122, 123, 124, 125.$$

We will now give a constructive proof for the existence of the greatest common divisor of two integers. This proof leads to an algorithm for computing the greatest common divisor called the **Euclidean** algorithm.

First, we will describe the steps in this algorithm and then we will verify that it does indeed compute the greatest common divisor.

Given two integers a and b, divide b by a with remainder via the Division algorithm, i.e. $b = aq + r_0$ with $0 \leq r_0 < a$. Now divide a by r_0, i.e. $a = r_0 q_0 + r_1$ with $0 \leq r_1 < r_0$. Repeat this process dividing r_0 by r_1 yielding remainder r_2, etc. In general, we divide r_k by r_{k+1} yielding remainder r_{k+2}. Now since $0 \leq r_{k+2} < r_{k+1} < \cdots < r_0$ and $r_0, r_1, \ldots, r_{k+2}$, etc. are all positive integers, eventually we must obtain a remainder of 0. We claim the remainder just prior to obtaining a remainder of zero is the $\gcd(a, b)$.

Example 3.4 *Before we demonstrate this proof that we have indeed obtained the greatest common divisor via the Euclidean algorithm, let's first illustrate it with an example. We shall compute* $\gcd(252, 270)$.

$$270 = 252(1) + 18 \quad with \quad 0 \leq r_0 = 18 < 252.$$

$$252 = 18(14) + 0 \quad so \quad r_1 = 0 \quad and \quad \gcd(252, 270) = 18.$$

Now we could have first divided 252 by 270 to get

$$252 = 270(-1) + 18 \quad with \quad 0 \leq r_0 = 18 < 270.$$

$$270 = 18(15) + 0,$$

and so we arrive at the same result.

Example 3.5 *We will compute* $\gcd(81, 57)$ *using the Euclidean algorithm.*

$$81 = 57(1) + 24 \quad with \quad 0 \leq r_0 = 24 < 57.$$

$$57 = 24(2) + 9 \quad with \quad 0 \leq r_1 = 9 < 24.$$

$$24 = 9(2) + 6 \quad with \quad 0 \leq r_2 = 6 < 9.$$

$$9 = 6(1) + 3 \quad with \quad 0 \leq r_3 = 3 < 6.$$

$$6 = 3(2) + 0 \quad so \quad r_4 = 0 \quad and \quad \gcd(81, 57) = r_3 = 3.$$

Proof 3.5 *Now we prove that the Euclidean algorithm does indeed produce the greatest common divisor of two integers. For this we need to prove the result satisfies the three parts of the definition of the greatest common divisor. Let r_k be the last non-zero remainder produced by the Euclidean algorithm. Certainly, $r_k > 0$. we now show r_k divides both a and b. Since $r_{k+1} = 0$, we know the Division algorithm yielded $r_{k-1} = q_k r_k$. Therefore $r_k \mid r_{k-1}$. Since $r_{k-2} = q_{k-1}r_{k-1} + r_k$ and r_k divides both r_{k-1} and r_k, it must also divide r_{k-2} (see Exercise 5 in Section 2.1). Continuing this way, r_k divides r_{k-1} and r_{k-2}, etc. until we arrive at $r_k \mid r_1$ and $r_k \mid r_0$. Since $a = r_0 q_0 + r_1$ we have $r_k \mid a$ and thus since $b = aq + r_0$ we also have $r_k \mid b$.*

Now suppose we have another common divisor d with $d \mid a$ and $d \mid b$. Since $r_0 = b + a(-q)$ this implies $d \mid r_0$. Since $r_1 = a + r_0(-q_0)$ we also have $d \mid r_1$. Continuing in this manner we eventually have $d \mid r_{k-2}$ and $d \mid r_{k-1}$. Since $r_k = r_{k-2} + r_{k-1}(-q_k)$, we then get $d \mid r_k$. □

One consequence of the non-constructive proof for the existence of the greatest common divisor of two integers a and b was that there exists integers x_0 and y_0 such that $\gcd(a, b) = ax_0 + by_0$. We say that the greatest common divisor of a and b can be represented as a **linear combination** of a and b. By using the steps in the Euclidean algorithm, we can derive these integers x_0 and y_0 in a constructive way called the **Extended Euclidean** algorithm. We will illustrate the algorthm in an example.

Example 3.6 *Refer to the Euclidean algorithm worked out for Example 3.5. Set $a = 57$ and $b = 81$. Note that $r_3 = 3 = \gcd(a, b)$. Let's list the steps of Example 3.5 in reverse order.*

$$r_1 = r_2(1) + r_3.$$

$$r_0 = r_1(2) + r_2.$$

$$a = r_0(2) + r_1.$$

$$b = a(1) + r_0.$$

Then

$$r_3 = (-1)r_2 + r_1.$$

$$r_2 = (-2)r_1 + r_0 \quad \textit{so by sustituting}$$
$$r_3 = (-1)[(-2)r_1 + r_0] + r_1 = (3)r_1 + (-1)r_0.$$

$$r_1 = (-2)r_0 + a \quad \textit{so by sustituting}$$
$$r_3 = (3)[(-2)r_0 + a] + (-1)r_0 = (-7)r_0 + (3)a.$$

$$r_0 = (-1)a + b \quad \textit{so by sustituting}$$
$$r_3 = (-7)[(-1)a + b] + (3)a = (10)a + (-7)b.$$

In other words, $x_0 = 10$, $y_0 = -7$ and $\gcd(57, 81) = (10)(57) + (-7)(81)$.

Example 3.7 *Assuming the result that we have prime factorization of natural numbers as in Theorem 2.4 of Section 2.5, there then is another simple and semi-constructive way of computing the greatest common divisor of two positive integers. If $a = p_1^{e_1} p_2^{e_2} \cdots p_n^{e_n}$ and $b = p_1^{f_1} p_2^{f_2} \cdots p_n^{f_n}$ where $p_1, p_2, \ldots, p_n$ are prime numbers and the e_i's and f_i's are natural numbers, then*

$$\gcd(a, b) = p_1^{\min(e_1, f_1)} p_2^{\min(e_2, f_2)} \cdots p_n^{\min(e_n, f_n)}.$$

Let's illustrate the result with an example. We will show that $\gcd(252, 270) = 18$ as was computed in Example 3.4. First, note that

$$252 = 2^2 \cdot 3^2 \cdot 7 \quad \textit{and} \quad 270 = 2 \cdot 3^3 \cdot 5.$$

However, to make avail of our result we need to fill in the gaps with the missing primes. In other words,

$$252 = 2^2 3^2 5^0 7^1 \quad \textit{and} \quad 270 = 2^1 3^3 5^1 7^0.$$

Then

$$\gcd(252, 270) = 2^{\min(2,1)} 3^{\min(2,3)} 5^{\min(0,1)} 7^{\min(1,0)} = 2^1 3^2 5^0 7^0 = 18.$$

We need to point out that this method for finding the greatest common divisor is still non-constructive, since the proof of Theorem 2.4 for factoring a number into prime numbers was non-constructive. Indeed, factoring a number into prime numbers is a computationally intensive exercise and is the basis for some cryptographic encryption algorithms.

EXERCISES

1 Give a constructive proof that every odd integer is a sum of two consecutive integers and apply your result to the odd integer 251.

2 Give a constructive proof that every odd integer is a difference of two consecutive perfect squares and apply your result to the odd integer 251.

3 Give a (very short) constructive existence proof (with two cases) of the following:

$$\forall x \in \mathbb{R}^{>0} \; \exists n \in \mathbb{N} \text{ such that } n > x.$$

a. Now show $\forall x, y \in \mathbb{R}^{>0} \; \exists n \in \mathbb{N}$ such that $nx > y$.

b. Now show $\forall x \in \mathbb{R}^{>0} \; \exists n \in \mathbb{N}$ such that $\frac{1}{n} < x$.

4 We give a constructive proof that there exists two irrational numbers a and b such that a^b is rational. This is done by exhibiting a specific example: Consider $a = \sqrt{2}$ and $b = \log_2 9$.

a. Prove by contradiction that b is irrational.

b. Verify directly that a^b is rational.

5 Compute $\gcd(1071, 462)$ in two different ways.

a. Using the Euclidean Algorithm.

b. Using the result in Example 3.7.

6 Using the Extended Euclidean algorithm, express $\gcd(1071, 462)$ as a linear combination of 1071 and 462.

7 Consider the natural numbers $a = 126$ and $b = 35$.

a. Use the Euclidean Algorithm to find $d = \gcd(126, 35)$.

b. Use the Extended Euclidean Algorithm to find $x_0, y_0 \in \mathbb{Z}$ such that $126x_0 + 35y_0 = d$.

8 Consider the natural numbers $a = 83$ and $b = 38$.

    a. Use the Euclidean Algorithm to find $d = gcd(83, 38)$.

    b. Use the Extended Euclidean Algorithm to find $x_0, y_0 \in \mathbb{Z}$ such that $83x_0 + 38y_0 = d$.

9 Prove that the statement made in Example 3.7 does indeed yield the greatest common divisor of two positive integers.

10 Let a and b be two positive integers. The integer l is the **least common multiple** of a and b, written $l = \mathrm{lcm}(a, b)$ if

    a. $l > 0$

    b. $a|l$ and $b|l$ (common multiple) and

    c. $a|m$ and $b|m$ implies $l|m$ (least).

One can prove that $\mathrm{lcm}(a, b)$ exists by verifying that if $m = p_1^{e_1} p_2^{e_2} \cdots p_n^{e_n}$ and $n = p_1^{f_1} p_2^{f_2} \cdots p_n^{f_n}$ where $p_1, p_2, \ldots, p_n$ are prime numbers and the e_i's and f_i's are natural numbers, then

$$\mathrm{lcm}(m, n) = p_1^{\max(e_1, f_1)} p_2^{\max(e_2, f_2)} \cdots p_n^{\max(e_n, f_n)}.$$

    a. Verify that the above statement about the least common multiple is indeed True with a proof.

    b. Using the above statement compute $\mathrm{lcm}(252, 270)$.

11 Prove for positive integers a and b that

$$ab = \gcd(a, b)\mathrm{lcm}(a, b).$$

12 Give a non-constructive proof for the following result: If a and b are positive real numbers with $b - a > 1$, then there exists a natural number m such that $a < m < b$. Prove this by considering the set $\{n \in \mathbb{N} : a < n\}$.

13 As a follow up to the previous exercise, show that for every positive real number $a \notin \mathbb{N}$ that there exists a natural number m such that $a < m < a + 1$.

3.2 UNIQUENESS PROOFS

What often goes hand-in-hand with an existence proof is a uniqueness proof. In other words, having shown $\exists x\ P(x)$, we now wish to show $\exists! x\ P(x)$, i.e. that this x which illustrates $P(x)$ is in fact unique. There are two ways of going about proving a uniqueness result, and it typically depends on whether your existence proof was constructive or non-constructive.

Uniqueness Proof Approaches: When proving the statement $\exists! x\ P(x)$,

1. If the proof of $\exists x\ P(x)$ was constructive, i.e. we constructed an x_0 such that $P(x_0)$ is True, one can show that if any x_1 should satisfy $P(x)$, then $x_1 = x_0$.

2. One can show that if x_1 and x_2 satisfy $P(x)$, then $x_1 = x_2$.

Example 3.8 *We start with a very simple example of the first approach. We shall prove that for any real numbers a and b with $a \neq 0$, the algebraic equation $ax = b$ has a unique solution.*

Proof 3.6 *For the existence proof we construct the solution $x_0 = \frac{b}{a}$, which of course is a solution to $ax = b$, since $a\left(\frac{b}{a}\right) = b$.*

For the uniqueness proof, suppose that x_1 is a solution to $ax = b$, i.e. $ax_1 = b$. Then dividing both sides by $a \neq 0$ we get $x_1 = \frac{b}{a} = x_0$. □

Example 3.9 *In this example we prove the uniqueness of the greatest common divisor using the second approach. We have already given a non-constructive proof for its existence.*

Proof 3.7 *To show uniqueness, assume d and d' are both greatest common divisors of a and b. Since $d|a$ and $d|b$ and $d' = \gcd(a,b)$, by part 3 of the definition of greatest common divisor, this implies $d|d'$. Now reversing roles and using the same argument, since $d'|a$ and $d'|b$ and $d = \gcd(a,b)$, we have $d'|d$. Now by Exercise 5c. of Section 2.1, we have $d = \pm d'$. However $d, d' > 0$, so we can conclude that $d = d'$.* □

This example proves the uniqueness of q and r in the Division algorithm using the second approach, and it uses a proof by contradiction. We have already given a non-constructive proof of their existence.

Proof 3.8 *To show uniqueness of q and r, suppose that $n = qd + r$ and $n = q'd + r'$ with $0 \leq r, r' < d$. Without loss of generality, assume that $q \geq q'$. Suppose, to the contrary, that $q > q'$ and so as integers we would have $q \geq q' + 1$. Then $r' = n - q'd \geq n - (q-1)d = r + d \geq d$, a contradiction. Hence $q = q'$ and so $r = n - qd = n - q'd = r'$ thus proving uniqueness.* □

Example 3.10 *In this example we prove the uniqueness part of the Fundamental Theorem of Arithmetic, namely that every natural number ≥ 2 is either prime or can be written uniquely as a product of primes. This again uses the second approach for proving uniqueness and with a prooof by contradiction.*

Proof 3.9 *To prove uniqueness, suppose to the contrary that there are integers greater than 1 which have more than one prime factorization and set U equal to the set of all such integers. Since U is non-empty, let m be the smallest element of U. Say $m = p_1^{e_1} p_2^{e_2} \cdots p_n^{e_n}$ and $m = q_1^{f_1} q_2^{f_2} \cdots q_k^{f_k}$ where $p_1 < p_2 < \cdots p_n$ and $q_1 < q_2 < \cdots q_k$ are primes and $e_1, e_2, \ldots, e_n$ and $f_1, f_2, \ldots, f_k$ are positive integers. Equating these two factorizations we see that p_1 divides $q_1^{f_1} q_2^{f_2} \cdots q_k^{f_k}$ and by Corollary 4.2, $p_1 | q_i$ for some i, $1 \leq i \leq k$. But then $p_1 = q_i$, and so by cancellation we have $p_1^{e_1 - 1} p_2^{e_2} \cdots p_n^{e_n} = q_1^{f_1} \cdots q_i^{f_i - 1} \cdots q_k^{f_k}$. Set $r = p_1^{e_1 - 1} p_2^{e_2} \cdots p_n^{e_n} = q_1^{f_1} \cdots q_i^{f_i - 1} \cdots q_k^{f_k}$. Since $r < m$ we know that r has a unique prime factorization, and so $p_1^{e_1 - 1} p_2^{e_2} \cdots p_n^{e_n}$ and $q_1^{f_1} \cdots q_i^{f_i - 1} \cdots q_k^{f_k}$ must be identical factorizations. If we throw $p_1 = q_i$ back in to get the two prime factorizations of m, then $p_1^{e_1} p_2^{e_2} \cdots p_n^{e_n}$ and $q_1^{f_1} q_2^{f_2} \cdots q_k^{f_k}$ must be the same as well, which is a contradiction to $m \in U$.* □

EXERCISES

1 Give the uniqueness proof for Exercise 1 in Section 3.1 that every odd integer is a sum of two consecutive integers and apply your result to the odd integer 251.

2 Give the uniqueness proof for Exercise 2 in Section 3.1 that every odd integer is a difference of two consecutive perfect squares and apply your result to the odd integer 251.

3 Prove that for every acute angle α there exists a unique obtuse angle β such that $\alpha + \beta = \pi$.

4 Show for every positive real number $a \notin \mathbb{N}$ there exists a unique natural number m such that $a < m < a + 1$. (see Exercise 13 in Section 3.1)

5 Prove only the uniqueness of the following result: For every positive real number $a \notin \mathbb{N}$ there exists a unique natural number m such that $|a - m| < \frac{1}{2}$ (you will need the triangle inequality – Exercise 1d. of Section 2.6).

CHAPTER 4

Foundational Mathematical Topics

I N THIS CHAPTER we present a collection of topics which serve as a *proving ground* (pun intended) for practicing the techniques you have studied in the first three chapters. These topics are prevalent in mathematics and thus also serve as a foundation for more in-depth mathematical pursuits. In Section 4.1, we present set relations, and in particular equivalence relations which foreshadow many important structures in abstract mathematics. In Section 4.2, we present basic concepts revolving around the topic of functions, and many areas of mathematics will have special functions associated with them. Section 4.3 will highlight some fundamental properties of number theory which play a role in many fields of mathematics. In Section 4.4, we describe in more detail congruence modulo n which is an equivalence relation with additional properties and another fundamental topic in mathematics. In the last two sections we dig deeper into two particular subjects. In Section 4.5 we introduce the reader to the notion that there are different sizes of infinity in mathematics, and in Section 4.6 we present a wonderful counting argument that requires some impressive theoretical mathematics and technical tools.

4.1 SET RELATIONS

The notion of a relation on a set is important in many fields of mathematics. There are many applications of a particular type of relation (called an equivalence relation) which you will encounter as you study deeper topics in mathematics. We start by defining the Cartesian product, then define a relation, and finally narrow things down to an equivalence relation.

DOI: 10.1201/9781032687728-4

In the process of introducing the Cartesian product we add a nice combinatorial result called the Inclusion-Exclusion Principle.

4.1.1 Cartesian Product

Sets do not have a sense of ordering. Indeed, for instance the set $\{a, b\}$ is equal to the set $\{b, a\}$. So one can think of this set as an *unordered* pair of elements. We need an object, let's denote it by (a, b) which has the property that $(a, b) \neq (b, a)$ which we will call an **ordered pair** of elements. For intuition, think of the Cartesian coordinate system. The point $(1, 0)$ is in a different spot in two-dimensional space from the point $(0, 1)$. Now we could naively define it to have this property that $(a, b) \neq (b, a)$ or we could be more formal and define it in terms of sets, because ultimately mathematics is based on sets. In fact, the formal definition of (a, b) is $\{\{a\}, \{a, b\}\}$ so that certainly, for $a \neq b$, $\{\{a\}, \{a, b\}\} \neq \{\{b\}, \{b, a\}\}$, since the first contains a singleton set $\{a\}$ while the second contains a singleton set $\{b\}$. We are now ready to define the Cartesian product.

One final bit of terminology before we do so. The element a will be called the **first coordinate** of (a, b) and b will be called the **second coordinate**.

Definition 4.1 *Given any two sets A and B, the* **Cartesian product** *of A and B, written*

$$A \times B = \{(a, b) \ : \ a \in A \ and \ b \in B\}.$$

If $A = B$ we sometimes write $A \times A$ as A^2 with the formal notation of $\times$ extended to express exponentiation.

Example 4.1 *Here we give several examples to illustrate the definition of Cartesian product.*

1. *Let $A = \{a, b, c\}$ and $B = \{d, e\}$. Then*

$$A \times B = \{(a, d), (a, e), (b, d), (b, e), (c, d), (c, e)\}.$$

2. *The Cartesian coordinate system from analytic (coordinate) geometry can be represented as $\mathbb{R}^2$.*

3. *The Cartesian product $\mathbb{Z} \times \mathbb{Z}$, or $\mathbb{Z}^2$, represents the discrete set of integer coordinate points in $\mathbb{R}^2$ (see Figure 4.1).*

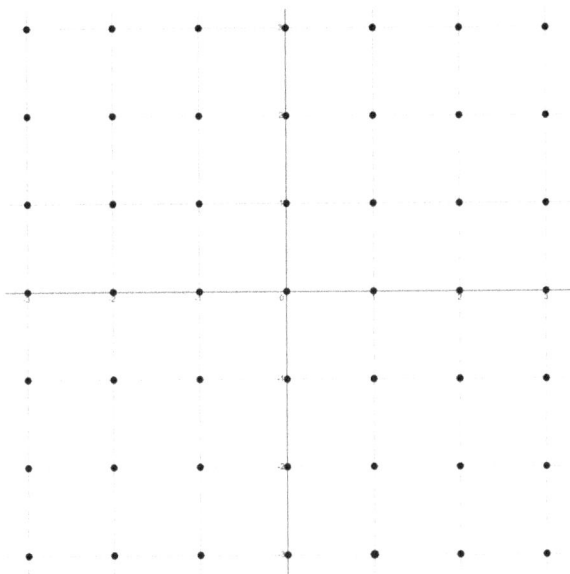

Figure 4.1 $\mathbb{Z} \times \mathbb{Z}$ or $\mathbb{Z}^2$

4. Let $A = \{x \in \mathbb{R} \mid 1 \leq x \leq 2\}$ and $B = \{x \in \mathbb{R} \mid 1 < x \leq 2\}$. Each of these sets is called an **interval** of real numbers. There is notation for such intervals. For instance, $A = [1, 2]$ and $B = (1, 2]$. Figure 4.2 illustrates what $A \times B$ looks like in $\mathbb{R}^2$.

We now present some basic set theoretic properties of the Cartesian product.

Proposition 4.1 *For any sets A, B, C and D,*

1. $A \times (B \cap C) = (A \times B) \cap (A \times C)$.

2. $A \times (B \cup C) = (A \times B) \cup (A \times C)$.

3. $(A \times B) \cap (C \times D) = (A \cap C) \times (B \cap D)$.

4. $(A \times B) \cup (C \times D) \subseteq (A \cup C) \times (B \cup D)$.

Proof 4.1 *We will prove the first and third statement and leave the rest as an exercise for the reader. For the first statement, Figure 4.3 represents an illustrative sketch of what we are trying to prove. It is not meant to substitute for a formal proof, however such pictures can*

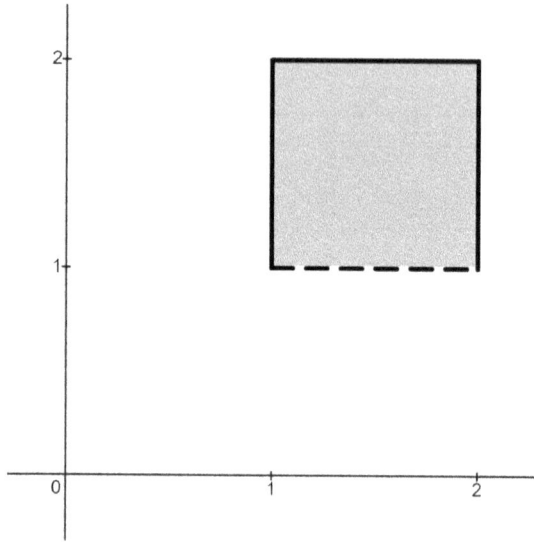

Figure 4.2 $[1, 2] \times (1, 2]$ consists of all points in the grey region and excluding the lower boundary.

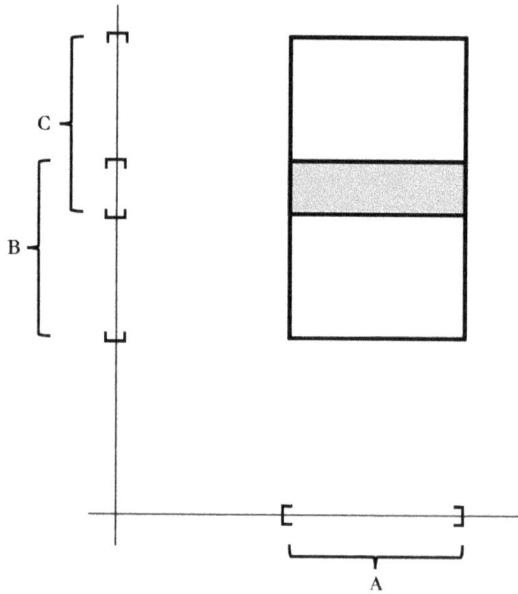

Figure 4.3 An illustrative picture of the first statement of Proposition 4.3.1. There area indicated in grey is the resulting set in question.

sometimes convince us of the truth of the statement and sometimes give hints on how to attack the proof.

Now $(x, y) \in A \times (B \cap C)$ iff $(x \in A) \wedge (y \in B \cap C)$ iff $(x \in A) \wedge (y \in B \wedge y \in C)$. Now by Exercise 4c. of Section 1.1, this last statement is logically equivalent to $(x \in A \wedge y \in B) \wedge (x \in A \wedge y \in C)$ which is True iff $(x, y) \in A \times B \wedge (x, y) \in A \times C$ iff $(x, y) \in (A \times B) \cap (A \times C)$.

For the third statement, $(x, y) \in (A \times B) \cap (C \times D)$ iff $(x, y) \in A \times B \wedge (x, y) \in C \times D$ iff $(x \in A) \wedge (y \in B) \wedge (x \in C) \wedge (y \in D)$ iff $(x \in A) \wedge (x \in C) \wedge (y \in B) \wedge (y \in D)$ iff $(x \in A \cap C) \wedge (y \in B \cap D)$ iff $(x, y) \in (A \cap C) \times (B \cap D)$. □

We will now give several results regarding the size of certain sets with the ultimate goal of counting the size of the Cartesian product. We give more than is required for this result, however these results are important and worth noting down here in the text. These counting results fall under a wonderful area of mathematics called **combinatorics**.

Theorem 4.1 *Let A and B be finite sets.*

1. *If A and B are disjoint, then $|A \cup B| = |A| + |B|$.*

2. *In general, $|A \cup B| = |A| + |B| - |A \cap B|$.*

Proof 4.2 *The first statement is self-evident. Indeed, if two sets have nothing in common, then to count the union, simple count each set separately and take the sum.*

For the second statement, using Exercise 4b. in Section 2.2 , express

$$A \cup B = (A - B) \cup (B - A) \cup (A \cap B) = (A \cap B') \cup (B \cap A') \cup (A \cap B).$$

We leave it as an exercise to verify that these last three sets are pairwise disjoint. Note also, using Exercise 4c. in Section 2.2, that

$$A = (A \cap B') \cup (A \cap B) \quad and \quad B = (B \cap A') \cup (A \cap B).$$

Therefore,

$$|A| + |B| = |A \cap B'| + |A \cap B| + |B \cap A'| + |A \cap B|$$

$$= |(A \cap B') \cup (B \cap A') \cup (A \cap B)| + |A \cap B| = |A \cup B| + |A \cap B|.$$

Solving for $|A \cup B|$ *in this last string of equations, we have*

$$|A \cup B| = |A| + |B| - |A \cap B|.$$

□

Now there is a more general result for counting the size of the union of any number of finite sets, called the **Inclusion-Exclusion Principle**, which we will shall now prove.

Theorem 4.2 (Inclusion-Exclusion Principle) *If* $A_1, A_2, \ldots, A_n$ *are finite sets where* $n \geq 2$, *then*

$$|A_1 \cup A_2 \cup \cdots \cup A_n| = \sum_{i=1}^{n} |A_i| - \sum_{1 \leq i < j \leq n} |A_i \cap A_j|$$

$$+ \sum_{1 \leq i < j < k \leq n} |A_i \cap A_j \cap A_k| + \cdots + (-1)^n |A_1 \cap A_2 \cap \cdots \cap A_n|.$$

Proof 4.3 *We will demonstrate this result using Weak Induction on* n, *the number of sets. The base case of* $n = 2$ *follows immediately from Theorem 4.1.2. For the inductive step (show* $P(n-1) \to P(n)$ *is True) with* $n > 2$; *again using Theorem 4.1.2,*

$$|A_1 \cup A_2 \cup \cdots \cup A_n| = |(A_1 \cup A_2 \cup \cdots \cup A_{n-1}) \cup A_n|$$

$$= |A_1 \cup A_2 \cup \cdots \cup A_{n-1}| + |A_n| - |(A_1 \cup A_2 \cup \cdots \cup A_{n-1}) \cap A_n|.$$

Using Exercise 3 in Section 2.2, this in turn is equal to

$$|A_1 \cup A_2 \cup \cdots \cup A_{n-1}| + |A_n| - |(A_1 \cap A_n) \cup (A_2 \cap A_n) \cup \cdots \cup (A_{n-1} \cap A_n)|.$$

Using twice the inductive assumption for $n - 1$ sets, this in turn equals

$$\left[\sum_{i=1}^{n-1} |A_i| - \sum_{1 \le i < j \le n-1} |A_i \cap A_j| + \cdots + (-1)^{n-1}|A_1 \cap A_2 \cap \cdots \cap A_{n-1}| \right]$$

$$+ |A_n| - \left[\sum_{i=1}^{n-1} |A_i \cap A_n| - \sum_{1 \le i < j \le n-1} |A_i \cap A_j \cap A_n| + \cdots \right.$$

$$\left. + (-1)^{n-1}|A_1 \cap A_2 \cap \cdots \cap A_{n-1} \cap A_n| \right] = \sum_{i=1}^{n} |A_i| - \sum_{1 \le i < j \le n} |A_i \cap A_j|$$

$$+ \sum_{1 \le i < j < k \le n} |A_i \cap A_j \cap A_k| + \cdots + (-1)^{n-2}|A_1 \cap A_2 \cap \cdots \cap A_n|.$$

Finally, note that $(-1)^{n-2} = (-1)^n$. □

Example 4.2 *Using the Inclusion-Exclusion Principle, we will count the numbers from 1 up through 100 which are divisible by 2, 3, or 5. Define the following sets:*

$$A = \{k \in \mathbb{Z} \; : \; 0 < k \le 100 \text{ and } 2 \mid k\},$$

$$B = \{k \in \mathbb{Z} \; : \; 0 < k \le 100 \text{ and } 3 \mid k\}$$

and $C = \{k \in \mathbb{Z} \; : \; 0 < k \le 100 \text{ and } 5 \mid k\}.$

The set we wish to count is $A \cup B \cup C$, and it should be clear then that

$$A \cap B = \{k \in \mathbb{Z} \; : \; 0 < k \le 100 \text{ and } 6 \mid k\},$$

$$A \cap C = \{k \in \mathbb{Z} \; : \; 0 < k \le 100 \text{ and } 10 \mid k\},$$

$$B \cap C = \{k \in \mathbb{Z} \; : \; 0 < k \le 100 \text{ and } 15 \mid k\},$$

and $A \cap B \cap C = \{k \in \mathbb{Z} \; : \; 0 < k \le 100 \text{ and } 30 \mid k\}.$

Now all of these sets can be easily counted.

$$|A| = 100/2 = 50, \quad |B| = 99/3 = 33,$$

$$|C| = 100/5 = 20, \quad |A \cap B| = 96/6 = 16,$$

$$|A \cap C| = 100/10 = 10, \quad |B \cap C| = 90/15 = 6, \quad and$$

$$|A \cap B \cap C| = 90/30 = 3.$$

Therefore, using the Inclusion-Exclusion Principle with $n = 3$,

$$|A \cap B \cap C| = |A| + |B| + |C| - (|A \cap B| + |A \cap C| + |B \cap C|) + |A \cap B \cap C|$$

$$= 50 + 33 + 20 - (16 + 10 + 6) + 3 = 103 - 32 + 3 = 74.$$

Thus, 74 *of the numbers from* 1 *through* 100 *are divisible by* 2, 3 *or* 5 *which is quite a few!*

Proposition 4.2 *For any finite sets* A *and* B, $|A \times B| = |A||B|$.

Proof 4.4 *Set* $|A| = m$ *and* $|B| = n$. *Our proof will be demonstrated using Weak Induction on* n. *First, let's enumerate the elements in our sets.*

$$A = \{a_1, a_2, \ldots, a_m\} \quad and \quad B = \{b_1, b_2, \ldots, b_n\}.$$

By Proposition 4.1.2,

$$A \times B = A \times (\{b_1, b_2, \ldots, b_{n-1}\} \cup \{b_n\})$$
$$= (A \times \{b_1, b_2, \ldots, b_{n-1}\}) \cup (A \times \{b_n\}).$$

By induction,

$$|A \times \{b_1, b_2, \ldots, b_{n-1}\}| = |A||\{b_1, b_2, \ldots, b_{n-1}\}| = m(n-1).$$

Since,

$$A \times \{b_n\} = \{(a_1, b_n), (a_2, b_n), \ldots, (a_m, b_n)\}, \quad we \ have$$
$$|A \times \{b_n\}| = |A| = m.$$

Certainly the sets $A \times \{b_1, b_2, \ldots, b_{n-1}\}$ and $A \times \{b_n\}$ are disjoint sets, since elements in the first set have second coordinates which differ from the second coordinate b_n in the second set. Now, by Theorem 4.1.1,

$$|A \times B| = m(n-1) + m = m[(n-1)+1] = mn.$$

□

4.1.2 Relation

Before we define an equivalence relation, we will introduce a more general mathematical concept called a **relation**.

Definition 4.2 *A **relation** $\sim$ on a set A is simply any subset of the Cartesian product $A \times A$.*

If (a, b) is an element of $\sim$, instead of writing $(a, b) \in \sim$, we prefer to write $a \sim b$ and we say a **relates to** b.

Example 4.3 *Here we list a number of examples of relations.*

1. *Let $A = \{a, b, c, d\}$ and let $\sim$ be the subset $\{(a, b), (b, b), (c, d)\}$ of $A \times A$. For instance, according to our definition of $\sim$, we have $c \sim d$ or c relates to d.*

2. *Let $A = \mathbb{Z}$ and $\sim$ represent the relation $<$. For instance $(2, 5)$ is in $\sim$, or $2 \sim 5$, since $2 < 5$. In general, (n, m) is in $\sim$ or $n \sim m$ exactly when $n < m$.*

3. *Set $A = \mathcal{P}(\mathbb{Z})$, which as the reader recalls, represents all the subsets of $\mathbb{Z}$ and is called the power set of $\mathbb{Z}$. Let $\sim$ represent the relation $\subseteq$, i.e. subset. In other words, subset X and subset Y of $\mathbb{Z}$ relate exactly when $X \subseteq Y$.*

4. *Take any set A and let $\sim$ represent equality, i.e. $a \sim b$ exactly when $a = b$. In other words $\sim$ is the set $\{(a, a) : a \in A\}$.*

5. *Let $f : A \to B$ be a function from a set A to another set B. Define a relation on A as follows: $a \sim b$ iff $f(a) = f(b)$.*

6. *Let $A = \mathbb{Z}$ and define $\sim$ as follows: $n \sim m$ iff There exists an integer k such that $m = nk$. In other words, n divides m which we have expressed using the notation $n | m$. For instance, $(3, -15)$ is in $\sim$, since 3 divides -15 because $-15 = 3(-5)$.*

7. *Define a relation on the set $\mathbb{Z}$ as follows: Fix a positive integer n and define $m \sim k$ iff $n|(m - k)$. This relation is called* **congruence modulo n** *and in place of $m \sim k$ we typically write $m \equiv_n k$ or $m \equiv k \pmod{n}$. In this setting, the integer n is called the* **modulus**. *Section 4.4 is completely dedicated to this very important relation.*

8. *Define a relation on $\mathbb{Q}$ as follows: $\frac{a}{b} \sim \frac{c}{d}$ iff $ad = bc$. So for instance $(\frac{1}{2}, \frac{-3}{-6})$ is in $\sim$ or $\frac{1}{2} \sim \frac{-3}{-6}$, since $(1)(-6) = (2)(-3)$. In basic arithmetic, this relation is viewed as equality, but in reality hidden underneath this equality is a relation which we shall see is, in fact, an equivalence relation.*

There are various properties one may wish to investigate regarding a relation. We list a few below.

Definition 4.3 *Let $\sim$ be a relation on a set A. We say $\sim$ is*

1. **reflexive** *if for all $a \in A$ we have $a \sim a$.*

2. **symmetric** *if for all $a, b \in A$ we have $a \sim b$ implies $b \sim a$.*

3. **transitive** *if for all $a, b, c \in A$ we have $a \sim b$ and $b \sim c$ implies $a \sim c$.*

4. **irreflexive** *if for all $a \in A$ we have $a \nsim a$.*

5. **anti-symmetric** *if for all $a, b \in A$ we have $a \sim b$ and $b \sim a$ implies $a = b$.*

Example 4.4 *One can represent a relation by a directed graph. Consider the set $A = \{a, b, c, d\}$ and relation*

$$R = \{(a, a), (a, c), (b, a), (c, a), (c, b), (c, d), (d, b), (d, d)\}.$$

Figure 4.4 is the directed graph corresponding to R. The vertices of the graph are the elements of A and each relation in R is represented by an arrow connecting two vertices. The graph is called directed, since we connect vertices with arrows instead of simple edges. We need to make the graph directed, since relations do not have the symmetric property in general.

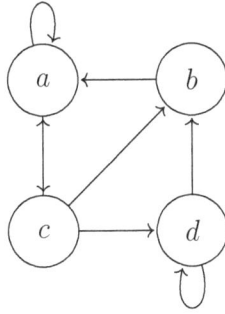

Figure 4.4 Directed graph representation of a relation.

Some examples of types of relations that are of particular importance in mathematics are the following:

Definition 4.4 *Let $\sim$ be a relation on a set A. We say that $\sim$ is a*

1. **partial ordering** *of A if it is reflexive, anti-symmetric and transitive, and in this case A is called a* **partially ordered** *set, or more briefly a* **poset**.

2. **equivalence relation** *on A if it is reflexive, symmetric and transitive.*

3. **function** *on A if for every $a \in A$ there exists an ordered pair (a, b) in $\sim$ for some $b \in A$, and this b is unique.*

Remark 4.1 *Here we make some remarks about these important relations just defined.*

1. *One can check that $\subseteq$ is a partial ordering on $\mathcal{P}(\mathbb{Z})$ and $\equiv_n$ is an equivalence relation on $\mathbb{Z}$, and if we restrict the relation* **divides** *to a relation on positive integers, then it becomes a partial ordering on positive integers.*

2. *For a partially ordered set not all elements must relate. For instance, in the case of the relation* **divides** *on positive integers, the numbers 2 and 3 do not relate, i.e. $2 \nmid 3$ and $3 \nmid 2$. There is a property called* **trichotomy** *which would rule out these occurrences. It says for any $a, b \in A$ and relation $\sim$, either $a \sim b$, $b \sim a$ or $a = b$. For instance, the relation $<$ on $\mathbb{Z}$ has the trichotomy property.*

Example 4.5 *We can represent partial orderings with a tree diagram, also called a* **Hasse diagram**. *Let's look at two examples.*

1. *Consider the subsets of the set $\{a, b, c\}$ partially ordered using $\subseteq$, i.e. subset. We can represent this ordering in a Hasse diagram as in Figure 4.5. In this type of diagram, elements below are related to elements above via a connection.*

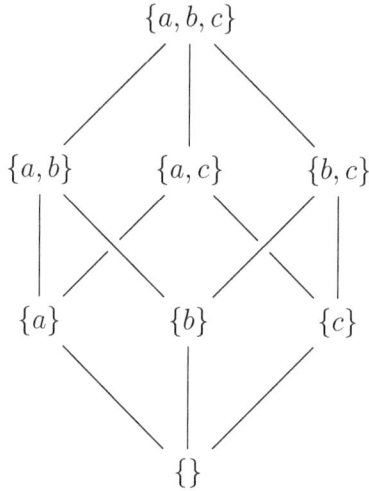

Figure 4.5 The Hasse diagram for subsets of $\{a, b, c\}$ partially ordered using $\subseteq$

2. *Let's consider the set of factors of 60 and the relation divides and create the Hasse diagram. In Figure 4.6, the numbers below divide*

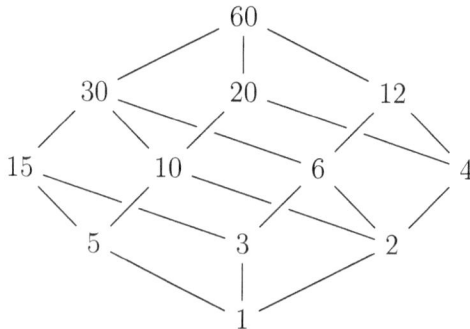

Figure 4.6 The Hasse diagram for factors of 60 partially ordered by divides

any of numbers above to which they are connected. In the construction of this Hasse diagram each level has the same number of prime factors.

There are many other orderings that mathematicians study. To name a few, for instance, there is a **Boolean algebra** and an **ordered lattice**. As you can tell, the idea of an ordering is a fundamental topic in mathematics.

4.1.3 Equivalence Relation

The focus of our discussion for the remainder of this section is equivalence relations. Let's list here the examples introduced already which are equivalence relations and verify the three properties.

Example 4.6 *Here are some examples of equivalence relations.*

1. *Take any set A and let $\sim$ be equality, i.e. $a \sim b$ exactly when $a = b$.*

   Reflexivity of equality states that $a = a$, which is self-evident. For symmetry, if $a = b$ then certainly $b = a$. Finally, for transitivity, if $a = b$ and $b = c$ it follows that $a = c$. Thus, equality is an equivalence relation. Indeed, the notion of an equivalence relation is in a sense a generalization of equality.

2. *Let $f : A \to B$ be a function from a set A to another set B. Define a relation on A as follows: $a \sim b$ iff $f(a) = f(b)$.*

   We leave the verification of this equivalence relation as an exercise.

3. *Define a relation on the set $\mathbb{Z}$ as follows: Fix a positive integer n and define $m \sim k$ iff $m \equiv k \pmod{n}$.*

   For reflexivity, we need to verify that $m \sim m$, or equivalently, $m \equiv m \pmod{n}$. This is equivalent to showing that $n \mid (m - m)$ or $n \mid 0$, but this last statement is True for any positive integer n, since $0 = n \cdot 0$.

   For symmetry, we assume $m \sim k$ and show $k \sim m$. If $m \sim k$, then $m \equiv k \pmod{n}$, which implies $n \mid (m - k)$. By definition, this implies there is an integer l such that $m - k = nl$. But then $(-1)(m - k) = (-1)nl$ or $k - m = n(-l)$ which says that $n \mid k - m$ or $k \equiv m \pmod{n}$, and thus $k \sim m$.

For transitivity, assuming $m \sim k$ and $k \sim l$ we show that $m \sim l$.
Thus, we are assuming that $m \equiv k \pmod n$ and $k \equiv l \pmod n$,
i.e. $n \mid (m - k)$ and $n \mid (k - l)$. We need to somehow get to the
statement $n \mid (m - l)$ so that $m \equiv l \pmod n$ and thus, $m \sim l$. We
get there using Exercise 5b. in Section 2.1. Indeed, $n \mid (m - k)$ and
$n \mid (k - l)$ implies $n \mid [(1)(m - k) + (1)(k - l)]$ which is the same
as saying $n \mid (m - l)$.

4. *Define a relation on $\mathbb{Q}$ as follows: $\frac{a}{b} \sim \frac{c}{d}$ iff $ad = bc$.*

   For reflexivity, $\frac{a}{b} \sim \frac{a}{b}$, since $ab = ba$.

   For symmetry, $\frac{a}{b} \sim \frac{c}{d}$ implies $ad = bc$ which implies $cb = da$, and
   so $\frac{c}{d} \sim \frac{a}{b}$.

   For transitivity, if $\frac{a}{b} \sim \frac{c}{d}$ and $\frac{c}{d} \sim \frac{e}{f}$, then $ad = bc$ and $cf = de$.
   We need to somehow get to the statement $af = be$ using these two
   equations, so that $\frac{a}{b} \sim \frac{e}{f}$. Here is how we do this. Notice that

   $$acf = a(cf) = a(de) = (ad)e = (bc)e = bce.$$

   Now cancel c from both sides of the equation $acf = bce$ and you
   have just what you need, namely $af = be$.

As we have stated already, for the remainder of this section we will
be assuming that $\sim$ is an equivalence relation and as such we typically
use the notation $\equiv$ in place of $\sim$. A very important set on the topic of
equivalence relations is the notion of an equivalence class. These classes
will come up again and again in mathematics.

Definition 4.5 *Let $\equiv$ be an equivalence relation on a set A and $a \in A$.*
*The **equivalence class** of a **with respect to** $\equiv$, written as*

$$[a]_\equiv \ \ equals \ \ \{ b \in A \ : \ a \equiv b\}.$$

*The element a is sometimes called a **representative** of the class $[a]_\equiv$.*
The collection of all equivalence classes of A with respect to $\equiv$, in other
*words $\{ [a]_\equiv \ : \ a \in A \}$, is denoted by $A/\equiv$ and is called the **quotient***
***set** of A.*

Thus, the quotient set is a set of sets. We shall see later on why it
gets the name **quotient**. At times we will simply write $[a]$ in place of

$[a]_\equiv$ when the equivalence relation is understood and we may simply call $[a]$ the **class of** a for brevity. Some other notation for an equivalence class which the reader may encounter here or in other texts is $\bar{a}$ in place of $[a]$.

Example 4.7 *Let's compute some equivalence classes for the examples already presented.*

1. *The equivalence classes for equality on a set A are singleton sets, i.e. $[a] = \{a\}$, since no other element besides a relates to a. Therefore, the quotient set $A/\equiv$ is $\{\, \{a\} \mid a \in A \}$.*

2. *For the equivalence relation we defined on $\mathbb{Q}$ (see Example 4.6.4) an equivalence class represents all the different ways we can represent a particular fraction. For instance, the equivalence class*

$$\left[\frac{1}{2}\right] = \left\{ \frac{1}{2}, \frac{-3}{-6}, \frac{12}{24}, \dots \right\}.$$

3. *Consider the equivalence relation congruence modulo 3 on $\mathbb{Z}$. There are exactly three distinct equivalence classes. Each class contains integers which when divided by 3 yield the same remainder.*

$$[0]_3 = \{0, \pm 3, \pm 6, \dots\}$$

$$[1]_3 = \{\dots, -8, -5, -2, 1, 4, 7, \dots\}$$

$$[2]_3 = \{\dots, -7, -4, -1, 2, 5, 8, \dots\}$$

Therefore, the quotient set $\mathbb{Z}/\equiv_3$ is $\{[0]_3,\ [1]_3,\ [2]_3\}$. Notice the notation $[a]_n$ which reminds us of what modulus n we are employing.

As already stated, one can view equivalence relations as a generalization of equality. Each class in a sense contains all the elements of a set which we view as being the same. Just consider the example of the equivalence class of $\frac{1}{2}$. In practice, we view $\frac{1}{2}$ and $\frac{-3}{-6}$ as being the same even though symbolically the look very different. Equivalence classes are simply a formal way of equating things which we wish to view as being equal.

We now prove a result which uncovers the essential properties of an equivalence relation.

Lemma 4.1 *Let $\equiv$ be an equivalence relation on a set A.*

1. *For all $a \in A$ we have $a \in [a]$.*

2. *For all $a, b \in A$ we have $[a] = [b]$ iff $a \equiv b$.*

3. *For all $a, b \in A$ either $[a] = [b]$ or $[a] \cap [b] = \emptyset$.*

Proof 4.5 *The first part follows immediately from the reflexive property. For the second part, assume first that $[a] = [b]$. Now since $a \in [a]$ we have $a \in [b]$ and so by definition and symmetry $a \equiv b$. Now assume that $a \equiv b$. Using transitivity and symmetry, notice that $c \in [a]$ iff $a \equiv c$ iff $c \equiv b$ iff $c \in [b]$ and so $[a] = [b]$. For the last part, either $[a] = [b]$ or $[a] \neq [b]$. In the latter case we show that $[a]$ and $[b]$ are disjoint, thus proving the result. Indeed, we prove this by proving the contrapositive. Therefore, assuming $[a] \cap [b] \neq \emptyset$, there is some $c \in [a] \cap [b]$. Then $c \in [a]$ and $c \in [b]$ and so $c \equiv a$ and $c \equiv b$. Using symmetry and transitivity we have $a \equiv b$ and so, by the second statement in this Lemma, $[a] = [b]$.* □

Notice that the second part of the lemma says that any element of a class can represent that class, i.e. if $b \in [a]$ then $[b] = [a]$. The first and third part of the lemma together say that equivalence classes divide the set A into a union of disjoint subsets which we call a partition of a set. Let's formally define this notion of a partition of a set.

Definition 4.6 *Let A be a non-empty set and $\mathcal{P}$ be a family of non-empty subsets of A. We say $\mathcal{P}$ is a **partition of** A or $\mathcal{P}$ **partitions** A if*

1. *For all $a \in A$ there is an $X \in \mathcal{P}$ such that $a \in X$.*

2. *For all $X, Y \in \mathcal{P}$ distinct we have $X \cap Y = \emptyset$.*

One can think of a partition of a set as a puzzle (see Figure 4.7) where each puzzle piece is an element of the partition and when you put all the puzzle pieces together you get the set A. According to this formal definition, we see from Lemma 4.1 that $A/\equiv$ is a partition of A.

Example 4.8 *Consider the earlier example of congruence modulo 3 an equivalence relation on $\mathbb{Z}$. The partition of $\mathbb{Z}$ into equivalence classes, namely $\mathbb{Z}/\equiv_3$, consists of three puzzle pieces, namely $[0]_3$, $[1]_3$ and $[2]_3$. These three classes are pairwise disjoint and their union is all of $\mathbb{Z}$.*

Figure 4.7 A partition is a puzzle.

EXERCISES

1 For the examples in Example 4.3, list three elements in each relation.

2 State and prove which of the properties in Definition 4.3 has each of the examples in Example 4.3.

3 Let A be the set of integers and $\sim$ be the relation $\leq$. State and prove which of the properties in Definition 4.3 this relation has.

4 Verify that $\subseteq$ is a partial ordering on $\mathcal{P}(\mathbb{Z})$ (you may use your work in Exercise 2).

5 Verify that if we restrict the relation **divides** to a relation on positive integers, then it becomes a partial ordering on positive integers.

6 Verify the relations defined in Example 4.6.2 is an equivalence relation (you may use your work in Exercise 2).

7 If $f : \mathbb{R} \to \mathbb{R}$ is the function defined by $f(x) = x^2$, describe the equivalence classes of Example 4.6.2.

8 For each of the following relations on $\mathbb{Z}$, decide if it's an equivalence relation. If it is, then verify; otherwise give a counter example.

   a. $a \sim b$ iff $|a - b| \leq 3$.

   b. $a \sim b$ iff $2|(a + b)$.

   c. $a \sim b$ iff $3|(a + b)$.

    d. $a \sim b$ iff $ab \leq 0$.

    e. $a \sim b$ iff $ab \geq 0$.

9 For each of the following relations on a set A, decide if it's an equivalence relation. If it is, then verify; otherwise give a counter example.

    a. A is the set of all English words. For $x, y \in A$ define $x \sim y$ iff x has a letter in common with y.

    b. $A = \mathbb{R}^2$ and for $(a, b), (c, d) \in A$ define $(a, b) \sim (c, d)$ iff $a + d = b + c$.

    c. Let $A = \{2, 3, 4, \ldots\}$. For $m, n \in A$ define $m \sim n$ iff there is a prime p such that $p \mid m$ and $p \mid n$.

10 Define the following relation on $\mathbb{R}^2$: $(a, b) \sim (c, d)$ iff $a^2 + b^2 = c^2 + d^2$.

    a. Verify that $\sim$ is an equivalence relation.

    b. Describe geometrically an equivalence class for $\sim$.

11 For each of the following relations, verify that it is an equivalence relation and then describe clearly a typical equivalence class.

    a. A is the set of all differentiable functions in one variable and $f \equiv g$ iff $f'(x) = g'(x)$.

    b. $A = \mathbb{R}^2$ and $(a, b) \equiv (c, d)$ iff $a = c$.

12 Suppose a relation $\sim$ on a set A has the following two properties:

    a. For all $a \in A$, $a \sim a$.

    b. For all $a, b, c \in A$ if $a \sim b$ and $b \sim c$, then $c \sim a$.

    Prove that $\sim$ is an equivalence relation on A.

13 Consider a circle divided into four equal sectors.

    a. If we can color each of the four sector either black or white, how many different colorings are there?

b. Let A be the set of colorings in part a, and define an relation as follows: Two colorings of the circle are equivalent if you can get from one to the other by rotating the circle by either $0°$, $90°$, $180°$, or $270°$. Prove that this relation is in fact an equivalence relation.

c. List the set of equivalence classes for this equivalence relation.

14 Using the formal definition of an ordered pair, i.e. $(a, b) = \{\{a\}, \{a, b\}\}$, prove that $(a, b) = (c, d)$ iff $a = c$ and $b = d$.

15 Prove the second and fourth statement in Proposition 4.1.

16 Referring to Proposition 4.1.4, give an example when $\subseteq$ is proper containment, i.e.

$$(A \times B) \cup (C \times D) \subsetneq (A \cup C) \times (B \cup D).$$

17 Prove that the three sets mentioned in the proof of Theorem 4.1.2 are pairwise disjoint.

18 Similar to Example 4.2, count the numbers from 1 up through 100 which are divisible by 2, 3, 5, or 7.

19 Prove the following statements:

- Given a partition $\mathcal{P}$ of a set A, the relation defined by $a \equiv b$ iff There is an $X \in \mathcal{P}$ such that $a, b \in X$ defines an equivalence relation whose equivalence classes consist precisely of the elements of $\mathcal{P}$.

- Conversely, if one starts with an equivalence relation $\equiv$ on a set A and forms the partition into equivalence classes and then defines an equivalence relation as we just did in the previous bullet, then we wind up with the same equivalence relation as we began with.

4.2 FUNCTIONS

In this section we present facts about functions that are fundamental and often used in mathematics. We begin with a reminder of some definitions and also present some new definitions and terminology.

Definition 4.7 *Let A and B be sets.*

1. The **Cartesian product** of A and B, written

$$A \times B = \{ (a,b) \ : \ a \in A \ \text{ and } \ b \in B \}.$$

2. A **relation** R from A to B is any subset of $A \times B$.

3. A **function** f from A to B, written $f : A \to B$, is a relation from A to B with the added properties that for each $a \in A$, $\exists! \, b \in B$ such that $(a,b) \in f$

4. Given a function f from A to B, if $(a,b) \in f$, then element $a \in A$ is called an **input** for f and the element $b \in B$ is called an **output** for f. We employ the notation $f(a) = b$ in this situation which basically says that if we input a into the function f, then the output is b.

5. Given a function f from A to B, the set A is called the **domain** of f and the set B is called the **codomain** of f. The **range** (or **image**) of f, written

$$f(A) = \{ b \in B \ : \ (a,b) \in f \}.$$

Notation for the domain of f is $D(f)$ and other notation for the range of f is $R(f)$.

6. Two functions f and g from A to B are **equal** if they are equal as sets. In other words, for all $a \in A$ we have $f(a) = g(a)$.

Example 4.9 *Here we present some specific examples of functions.*

1. Let $A = \{a, b, c\}$, $B = \{1, 2, 3\}$, and define $f : A \to B$ by

$$f = \{(a, 3), (b, 1), (c, 1)\}.$$

Then $f(a) = 3$, $f(b) = 1$, $f(c) = 1$, and $f(A) = \{1, 3\}$. In general, $f(A) \subseteq B$.

2. Functions can also be defined in terms of a formula. For instance, we can define $f : \mathbb{Z} \to \mathbb{Z}$ by $f(x) = x^2$. In which case we can generate the elements of f.

$$f = \{(0, 0), (1, 1), (-1, 1), (2, 4), (-2, 4), \ldots\}.$$

3. A **real-valued** *function has outputs which are real numbers. An example is*

$$f : \mathbb{N} \to \mathbb{R} \quad by \quad f(n) = \sqrt{n}.$$

4. *A real-valued function with domain $\mathbb{R}$ or $\mathbb{R}^2$ can be visually represented by its graph. Functions $f : \mathbb{R} \to \mathbb{R}$ have a graph which is a curve in $\mathbb{R}^2$ and satisfies the* **vertical line test**. *The vertical line test states that a curve in $\mathbb{R}^2$ represents a function if every vertical line passes through at most one point on the curve. In Figure 4.8,consider the curve that begins at the origin and climbs slowly from left to right. Three sample vertical lines are illustrated in the figure. Two pass through one point on the curve and one passes through none. It is clear that this curve passes the vertical line test. Indeed, this is, in fact, the graph of the function $f(x) = \sqrt{x}$.*

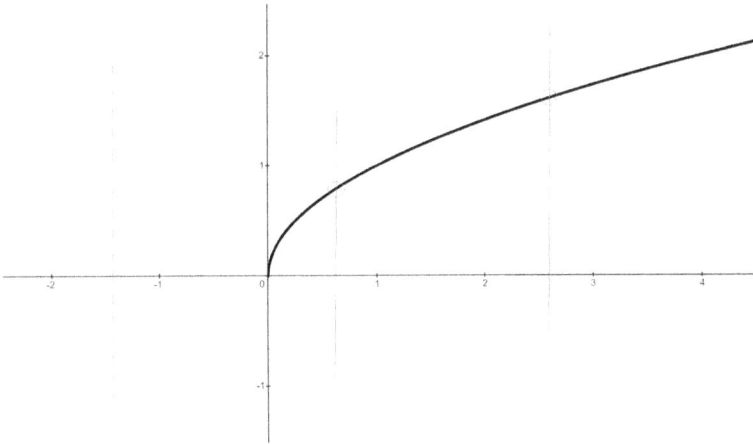

Figure 4.8 The Vertical Line Test.

A functions $f : \mathbb{R}^2 \to \mathbb{R}$ has a graph which is a surface in $\mathbb{R}^3$ and satisfies the **vertical line test**, *where vertical lines are lines parallel to the z-axis.*

Remark 4.2 *Here we add a few more remarks about functions.*

1. *We will move away from the ordered pair representation of a function, i.e. $(a, b) \in f$, and make use of the notation $f(a) = b$ for the most part.*

2. *Mathematicians sometimes refer to a function as a* **map**.

3. *We have seen that functions can be represented as a collection of ordered pairs or sometimes by a formula, and sometimes visually as a curve or surface depending on the setting. Another way to represent an arbitrary function, and this is sometimes seen in mathematical proofs, is with a simple sketch such as the one in Figure 4.9.*

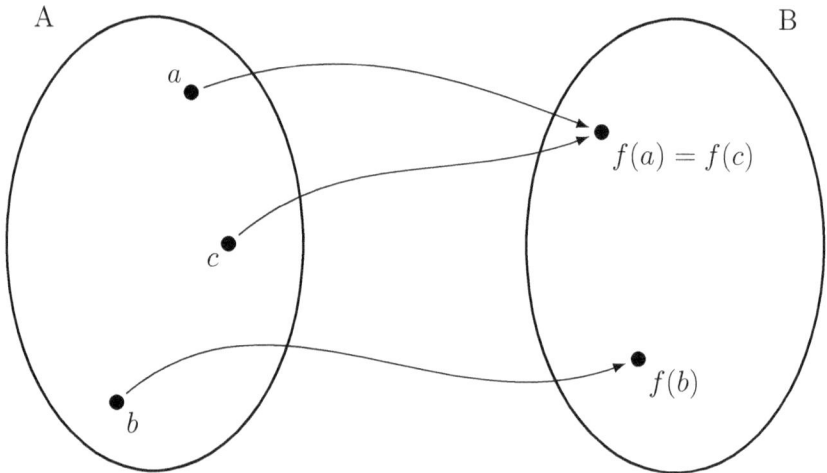

Figure 4.9 Visualization of a function.

Definition 4.8 *Let $f : A \to B$ be a function.*

1. *f is **injective** (or **one-to-one**) if for any $a_1, a_2 \in A$ whenever $f(a_1) = f(a_2)$ it must be the case that $a_1 = a_2$. In other words, in terms of the contrapositive statement, two different inputs must yield two different outputs. If f is represented by a curve in $\mathbb{R}^2$, an injective function satisfies the **horizontal line test**.*

2. *f is **surjective** (or **maps onto** B) if for every $b \in B$ there is an $a \in A$ such that $f(a) = b$. In other words, the image and codomain are equal.*

3. *f is a **bijection** if it is both injective and surjective. The function is also called a **one-to-one correspondence** between A and B, since every element in A is associated with a unique element in B and vice versa.*

Example 4.10 *These examples help illustrate the definitions just related and also demonstrate that there is no dependency between the two concepts of injective and surjective, i.e. neither property implies the other.*

1. *Let $f : \mathbb{N} \to \mathbb{N}$ by $f(n) = 2n$. This function is one-to-one, but does not map onto $\mathbb{N}$. Indeed, the image of f is the even numbers.*

2. *Let $f : \mathbb{N} \to \{0, 1\}$ by*

$$f(n) = \begin{cases} 0, & \text{if } n \text{ is even} \\ 1, & \text{if } n \text{ is odd} \end{cases}$$

This function is not one-to-one and maps onto $\{0, 1\}$.

3. *Let $f : \mathbb{N} \to \mathbb{N}$ by*

$$f(n) = \begin{cases} n/2, & \text{if } n \text{ is even} \\ n, & \text{if } n \text{ is odd} \end{cases}$$

*This function is not one-to-one and maps onto $\mathbb{N}$. To illustrate it not being one-to-one, notice that $f(1) = 1$ and $f(2) = 1$. This example and the previous one are examples of a **branch** or **piecewise defined** function. Such functions are defined differently on different parts of their domain.*

4. *Let $f : \mathbb{R} \to \mathbb{R}$ by $f(x) = 2x - 3$. This function is a bijection. Let's see why.*

   First, to prove f is one-to-one we will assume $f(a) = f(b)$, for any $a, b \in \mathbb{R}$, and we will show $a = b$. If $f(a) = f(b)$, then $2a - 3 = 2b - 3$. Adding 3 to both sides of the equation gives $2a = 2b$, and dividing both sides by 2 gets us to $a = b$, as desired.

   To prove f maps onto $\mathbb{R}$ we select an arbitrary real number b in the codomain, and attempt to find an input a such that $f(a) = b$. Equivalently, we need to solve for a in the equation $2a - 3 = b$. First add 3 to both sides to get $2a = b + 3$, and then divide both sides by 2 to get $a = \frac{b+3}{2}$. Since we could solve for a in general, this implies f maps onto $\mathbb{R}$. To see more concretely what a is for a particular instance, had we chosen $b = 5$, then $a = \frac{5+3}{2} = 4$. Indeed, $f(4) = 2(4) - 3 = 5$ as expected.

5. Let $f : \{0,1\} \to \{0,1\}$ by $f(0) = 0$ and $f(1) = 0$. This function is neither one-to-one nor maps onto $\{0,1\}$. Another way we could have represented f is as the set of ordered pairs $\{(0,0),(1,0)\}$.

Example 4.11 *We introduce some* **important** *examples of functions which occur in mathematical discourse.*

1. For any set A, the **identity map** on A, written $1_A : A \to A$ is defined by $1_A(a) = a$ for all $a \in A$. Certainly this map is a bijection.

2. If A is a proper subset of a set B, the **inclusion map** written $i : A \to B$ is defined by $i(a) = a$ for all $a \in A$. This map is again one-to-one, but certainly cannot map onto B. For a more concrete example, consider the map

$$i : \mathbb{Z} \to \mathbb{Q} \quad by \quad i(n) = \frac{n}{1}.$$

3. For any sets A and B, the **projection map** onto A, written $\pi_A : A \times B \to A$ is defined by $\pi_A(a,b) = a$. This function maps onto A but cannot be one-to-one if $|B| > 1$.

4. Consider any function $f : A \to B$. The **restriction map** of f to $C \subset A$, written $f \restriction C : C \to B$ is defined by $(f \restriction C)(c) = f(c)$ for all $c \in C$.

5. Consider any function $f : A \to B$. An **extension map** of f to $C \supset A$, written $\tilde{f} : C \to B$ is any map from C to B having the property that $\tilde{f}(a) = f(a)$ for all $a \in A$.

   A simple example of this is the following extension of $1_{\mathbb{Z}}$ to $\mathbb{Q}$ defined by

$$\tilde{1}_{\mathbb{Z}} : \mathbb{Q} \to \mathbb{Z} \quad by \quad \tilde{1}_{\mathbb{Z}}\left(\frac{m}{n}\right) = mn, \text{ assuming that } \frac{m}{n} \text{ is in lowest terms.}$$

6. Let $\equiv$ be an equivalence relation on a set A. The **quotient map** (or **canonical map**) $\nu : A \to A/\equiv$ is defined by $\nu(a) = [a]$. This function certainly maps onto $A/\equiv$.

7. Let A be any set. A function $\sigma : A \to A$ is a **permutation** of A if σ is a bijection. The collection of all permutations of A is denoted by $Sym(A)$ and is called the **symmetric group on** A.

When $A = \{1, 2, \ldots, n\}$, then $Sym(A)$ is typically denoted by S_n and is called the **symmetric group on** n. For a more concrete example, let's list all the elements of S_3 which we will denote by the set of bijections $\{\sigma_1, \sigma_2, \sigma_3, \sigma_4, \sigma_5, \sigma_6\}$.

$$\sigma_1(1) = 1, \quad \sigma_1(2) = 2, \quad \sigma_1(3) = 3.$$

$$\sigma_2(1) = 1, \quad \sigma_2(2) = 3, \quad \sigma_2(3) = 2.$$

$$\sigma_3(1) = 2, \quad \sigma_3(2) = 1, \quad \sigma_3(3) = 3.$$

$$\sigma_4(1) = 2, \quad \sigma_4(2) = 3, \quad \sigma_4(3) = 1.$$

$$\sigma_5(1) = 3, \quad \sigma_5(2) = 1, \quad \sigma_5(3) = 2.$$

$$\sigma_6(1) = 3, \quad \sigma_6(2) = 2, \quad \sigma_6(3) = 1.$$

Why are there exactly 6 elements? Can you come up with a formula for the size of S_n? We can interpret these bijections as operations on the vertices of a triangle where the vertices haves labels 1, 2, and 3 (Figure 4.10).

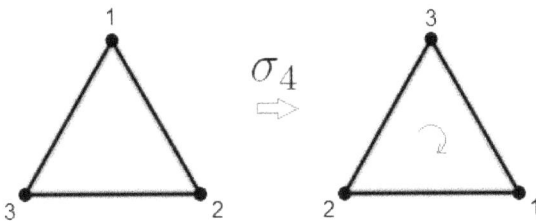

Figure 4.10 Clockwise rotation of the triangle by 120°

For instance, the bijection σ_4 rotates the triangle clockwise by 120°. Can you interpret what the other five bijections do? These bijections play an integral role in a field of mathematics called group theory.

Lemma 4.2 *Let $f : A \to B$ be a function.*

1. *f is a bijection   iff   For every $b \in B$ there exists a unique $a \in A$ such that $f(a) = b$.*

2. *If A and B are finite sets of the same size, then f is one-to-one iff f maps onto B.*

Proof 4.6 *We leave the proof of the first statement as an exercise. To prove the second statement, set $A = \{a_1, \ldots, a_n\}$ and $B = \{b_1, \ldots, b_n\}$. Then*

$$f(A) = \{f(a_1), \ldots, f(a_n)\} \subseteq B.$$

First assume that f is one-to-one. Then $f(a_1), \ldots, f(a_n)$ must be all distinct. Thus, $|f(A)| = n = |B|$ and so $f(A) = B$ proving f maps onto B. We prove the converse by contrapositive. If f is not one-to-one, then $f(a_1), \ldots, f(a_n)$ are not all distinct. Thus, $|f(A)| < n = |B|$ and so $f(A)$ is a proper subset of B proving f does not map onto B.

Definition 4.9 *Let $f : A \to B$ and $g : B \to C$. The **composition** function, written*

$$g \circ f = \{ (a, c) \ : \ \exists \, b \in B \text{ such that } (a, b) \in f \text{ and } (b, c) \in g \}.$$

In other words, $(g \circ f)(a) = g(f(a))$

Note that the domain of g is B so that $g(f(a))$ will always makes sense.

Example 4.12 *Here we give some examples of composition.*

1. *Consider the sets $A = \{a, b, c\}$, $B = \{0, 1, 2\}$, and $C = \{x, y, z\}$ and the functions*

$$f : A \to B \quad \text{defined by } f = \{(a, 0), (b, 0), (c, 2)\}.$$

$$g : B \to C \quad \text{defined by } g = \{(0, y), (1, z), (2, x)\}.$$

Then $g \circ f : A \to C$ is the function

$$g \circ f = \{(a, y), (b, y), (c, x)\}.$$

2. Let $f : \mathbb{R} \to \mathbb{R}$ and $g : \mathbb{R} \to \mathbb{R}$ be defined by $f(x) = x^2$ and $g(x) = \sqrt{x^2 + 1}$. Then

$$(g \circ f)(x) = g(f(x)) = g(x^2) = \sqrt{x^4 + 1}, \quad while$$

$$(f \circ g)(x) = f(g(x)) = f(\sqrt{x^2 + 1}) = \left(\sqrt{x^2 + 1}\right)^2 = x^2 + 1.$$

Notice that $(g \circ f)(1) = \sqrt{2}$ while $(f \circ g)(1) = 2$. Thus, in general when defined, composition is not commutative, i.e. $f \circ g \neq g \circ f$.

Definition 4.10 Let $f : A \to B$ be a function. The function $g : B \to A$ is an **inverse** of f if $g \circ f = 1_A$ and $f \circ g = 1_B$, i.e. for all $a \in A$ and for all $b \in B$,

$$g(f(a)) = a \quad and \quad f(g(b)) = b.$$

Example 4.13 Check for $f, g : \mathbb{R} \to \mathbb{R}$ defined by $f(x) = 2x - 3$ and $g(x) = \frac{x+3}{2}$ that g is an inverse of f.

In a sense, the inverse g of a function f *undoes* what the function f does. Below are some pertinent results concerning functions and their inverses:

Theorem 4.3 Let $f : A \longrightarrow B$ be a function from a set A to a set B.

1. f has an inverse iff f is one-to-one and maps onto B, i.e. f is a bijection.

2. If f has an inverse, then it has exactly one.

3. If f has an inverse, then the inverse is also one-to-one and maps onto A, i.e. the inverse is a bijection.

4. If $f_1 : A \to B$ has inverse g_1 and $f_2 : B \to C$ has inverse g_2, then $f_2 \circ f_1$ has an inverse, namely $g_1 \circ g_2$.

Proof 4.7 To prove the first statement, first we assume that f has an inverse g. We show f is one-to-one. For $a_1, a_2 \in A$, if $f(a_1) = f(a_2)$, then $g(f(a_1)) = g(f(a_2))$, i.e. $(g \circ f)(a_1) = (g \circ f)(a_2)$. By definition of *inverse, this equation reduces to $a_1 = a_2$, and we have proved that f is one-to-one. To show that f maps onto B, take any $b \in B$. We have to find an $a \in A$ such that $f(a) = b$. Set $a = g(b)$. Then*

$$f(a) = f(g(b)) = (f \circ g)(b) = b.$$

Now assume that f is a bijection. Define a function $g : B \longrightarrow A$ as follows: For $b \in B$, by Lemma 4.2, there is a unique $a \in A$ such that $f(a) = b$. We then define $g(b) = a$. Note that g is indeed a function, since the element $a \in A$ is uniquely determined. We now prove that this g is the inverse of f. First, for any $b \in B$, we have

$$(f \circ g)(b) = f(g(b)) = f(a) = b.$$

Second, take any $a \in A$. Set $b = f(a)$. Note that by definition of g, we have that $g(b) = a$. Hence,

$$(g \circ f)(a) = g(f(a)) = g(b) = a.$$

To prove the second statement, suppose that g_1 and g_2 are inverses of f. We will show that $g_1 = g_2$ and so f has only one inverse (when it exists). For any $b \in B$, since f is a bijection, by Lemma 4.2, there is a unique $a \in A$ such that $f(a) = b$. Then for all $b \in B$,

$$g_1(b) = g_1(f(a)) = (g_1 \circ f)(a) = a = (g_2 \circ f)(a) = g_2(f(a)) = g_2(b).$$

Hence, $g_1 = g_2$, since they are equal on every input. We leave the remaining proofs as exercises. □

Because of the fact that when an inverse exists there is only one, we can assign it notation without any confusion. The inverse of f will be denoted by f^{-1}. Take note that this is simply notation and should not be taken literally as $1/f$.

Example 4.14 *Here we give some examples of finding the inverse of a function.*

1. *If $A = \{a, b\}$, $B = \{c, d\}$ and $f : A \to B$ with $f = \{(a, c), (b, d)\}$, then f is a bijection, so f^{-1} exists with $f^{-1} = \{(c, a), (d, b)\}$. One simply reverses the order of each ordered pair in f.*

2. *Let's find the inverse mentioned in Example 4.13. In Example 4.10.4, we verified that f is one-to-one and maps onto $\mathbb{R}$, i.e. a bijection, and therefore has an inverse by Theorem 4.3. Indeed, in proving f maps onto $\mathbb{R}$ we found the formula $a = \frac{b+3}{2}$ which gives us the formula for the inverse, namely $f^{-1}(x) = \frac{x+3}{2}$.*

There is, however, a use of the notation f^{-1} which does not assume that the inverse of f exists.

Definition 4.11 *Let $f : A \to B$ be a function and $C \subseteq B$. The* **inverse image** *(or* **preimage**) *of C under f, written*

$$f^{-1}(C) = \{a \in A \ : \ f(a) \in C\}.$$

Example 4.15 *Let's compute a few examples of inverse image.*

1. *For Example 4.10.1, $f^{-1}(\{2\}) = \{1\}$, $f^{-1}(\{1\}) = \emptyset$, $f^{-1}(\{1,2\}) = \{1\}$ and $f^{-1}(\{1,2,3,4\}) = \{1,2\}$.*

2. *For Example 4.10.2, $f^{-1}(\{1\})$ equals the set of odd numbers.*

3. *For Example 4.10.3, $f^{-1}(\{2\}) = \{4\}$.*

4. *For Example 4.10.4, $f^{-1}(\{2\}) = \{5/2\}$.*

5. *For Example 4.10.5, $f^{-1}(\{1\}) = \emptyset$.*

4.2.1 Well-Definition

The last topic of this section deals with the notion of a **well-defined** map. This topic arises when an input can be represented in more than one way. In particular, we shall look at the situation when inputs are equivalence classes. Basically we want to check when such a map is indeed a function, i.e. If $a = b$ (two representations of the same input), then $f(a) = f(b)$. For otherwise, we would have a single input being sent to two different outputs contradicting the definition of a function. In our particular situation, we might have an equivalence relation $\equiv$ on a set A and $f : A/\equiv \to B$. We want to make sure f is well-defined, i.e. that f is by definition a function. This entails checking that whenever $[a_1] = [a_2]$ we have $f([a_1]) = f([a_2])$. This property is essential to many areas of abstract mathematics. In Section 4.4 we will see an important instance of well-definition.

Example 4.16 *We will illustrate this concept and its verification with several examples.*

1. *Consider the equivalence relation congruence modulo 3 on $\mathbb{Z}$. Define $f : \mathbb{Z}/\equiv_3 \to \mathbb{Z}$ by $f([n]) = n$. This map is **not** a well-defined function, since for instance $[0] = [3]$ while $f([0]) = 0 \neq 3 = f([3])$.*

2. *Consider the equivalence relation congruence modulo 2 and define $f : \mathbb{Z}/\equiv_2 \to \mathbb{Z}$ by*

$$f([n]) = \begin{cases} 0, & \text{if } n \text{ is even} \\ 1, & \text{if } n \text{ is odd} \end{cases}$$

We claim f is a well-defined function. If $[n] = [m]$, then $n \equiv_2 m$ and so $2|(n-m)$. Therefore, $n-m$ is even and so either n and m are both even or both odd. In either case, $f([n]) = f([m])$.

3. *Consider the equivalence relation $\equiv$ we defined earlier on $\mathbb{Q}$ (Example 4.6.4) and consider*

$$f : \mathbb{Q}/\equiv \to \mathbb{Q} \quad \text{defined by} \quad f([a/b]) = a^2/b^2.$$

We show that f is a well-defined function. If $[a/b] = [c/d]$, then $a/b \equiv c/d$ and so $ad = bc$. Using properties of $\mathbb{Z}$ we have $(ad)^2 = (bc)^2$ and so $a^2 d^2 = b^2 c^2$. Then $a^2/b^2 \equiv c^2/d^2$ which implies $f(a/b) = f(c/d)$.

EXERCISES

1 For each function in Example 4.10, verify the statements made for injective and surjective.

2 For each of the following functions, decide if it is injective, surjective or a bijection:

    a. $f : \mathbb{R}^{>0} \to \mathbb{R}^{>0}$ by $f(x) = \frac{1}{x}$.

    b. $f : \mathbb{R}^* \to \mathbb{R}^*$ by $f(x) = \frac{1}{x^2}$, where $\mathbb{R}^* = \mathbb{R} - \{0\}$

    c. $f : \mathbb{R} \to \mathbb{R}$ by $f(x) = \sin(x^2)$

    d. $f : \mathbb{Z} \times \mathbb{Z} \to \mathbb{Z}$ by $f(m, n) = 3m + n$.

e. $f : \mathbb{Z} \times \mathbb{Z} \to \mathbb{Z}$ by $f(m, n) = 4m + 2n$.

f. $f : \mathbb{Z} \times \mathbb{Z} \to \mathbb{Z} \times \mathbb{Z}$ by $f(m, n) = (2m, m + n)$.

g. $f : \mathbb{R} \times \mathbb{R} \to \mathbb{R} \times \mathbb{R}$ by $f(x, y) = (2x, x + y)$.

3 Carefully explain why the example given in Example 4.11.5 is indeed an extension.

4 Let $f : A \to B$, $g : B \to C$ and $h : C \to D$. Prove the following statements:

a. Composition is associative, i.e. $(h \circ g) \circ f = h \circ (g \circ f)$.

b. f, g one-to-one implies $g \circ f$ one-to-one.

c. If f maps onto B and g maps onto C, then $g \circ f$ maps onto C.

d. $g \circ f$ one-to-one implies f one-to-one.

e. $g \circ f$ maps onto C implies g maps onto C.

f. Give a counterexample to the following statement: $g \circ f$ one-to-one implies g one-to-one.

g. Give a counterexample to the following statement: $g \circ f$ maps onto C implies f maps onto B.

5 Let $A = \{a, b, c\}$ and $B = \{x, y, z\}$.

a. How many possible functions $f : A \to B$ are there?

b. How many possible surjective functions $f : A \to B$ are there?

c. How many possible injective functions $f : A \to B$ are there?

d. How many possible bijective functions $f : A \to B$ are there?

6 Consider the functions $f : A \to B$ and $g : B \to C$.

a. Prove that if f and g are bijections, then $g \circ f$ has an inverse and $(g \circ f)^{-1} = f^{-1} \circ g^{-1}$.

b. Give a concrete example where $g \circ f$ is a bijection, but f and g are not.

7 Prove Lemma 4.2.1

8 Prove the remaining parts of Lemma 4.3.

9 Compute $f^{-1}(\{1\})$ for Example 4.10.3.

10 Let $f : X \to Y$ be a function with $A, B \subseteq X$ and $C, D \subseteq Y$.

  a. Prove that $f(A \cup B) = f(A) \cup f(B)$.

  b. Give a counterexample to the statement $f(A \cap B) = f(A) \cap f(B)$.

  c. Prove that $f^{-1}(C \cup D) = f^{-1}(C) \cup f^{-1}(D)$.

  d. Prove that $f^{-1}(C \cap D) = f^{-1}(C) \cap f^{-1}(D)$.

11 Decide whether or not each of the following functions is well-defined:

  a. $f : \mathbb{Z}/\equiv_n \to \mathbb{Z}/\equiv_n$ by $f([a]) = [2a]$.

  b. $f : \mathbb{Z}/\equiv_n \to \mathbb{Z}/\equiv_n$ by $f([a]) = [ma + b]$, for some fixed integers m and b.

  c. $f : \mathbb{Z}/\equiv_n \to \mathbb{Z}/\equiv_n$ by $f([a]) = [a^2]$

  d. $f : \mathbb{Z}/\equiv_n \to \mathbb{Z}/\equiv_n$ by $f([a]) = [a^k]$, for some fixed positive integer k.

  e. $f : \mathbb{Z}/\equiv_2 \to \mathbb{Z}/\equiv_4$ by $f([a]_2) = [a]_4$.

  f. $f : \mathbb{Q} \to \mathbb{Z}$ by $f\left(\frac{m}{n}\right) = m$.

  g. $f : \mathbb{Z}/\equiv_4 \to \mathbb{Z}/\equiv_6$ by $f([x]_4) = [3x]_6$.

  h. $f : \mathbb{Q} \to \mathbb{Q}$ by $f\left(\frac{m}{n}\right) = \left(\frac{2m}{3n}\right)$.

  i. $f : \mathbb{Z}/\equiv_4 \to \mathbb{Z}/\equiv_6$ by $f([x]_4) = [x]_6$.

  j. $f : \mathbb{Z}/\equiv_4 \to \mathbb{Z}/\equiv_3$ by $f([m]_4) = [m^3]_3$.

  k. $g : \mathbb{Q} \to \mathbb{Q}$ by $g\left(\left[\frac{a}{b}\right]\right) = \left[\frac{a+b}{b}\right]$.

12 Consider $f : \mathbb{Z}/\equiv_{35} \to \mathbb{Z}/\equiv_5 \times \mathbb{Z}/\equiv_7$ by $f([x]_{35}) = ([x]_5, [x]_7)$.

  a. Prove f is a well-defined function.

  b. Is f is injective? (justify)

  c. Is f is surjective? (justify)

    d. Is f is bijective? (justify)

    e. Does f have an inverse? (justify)

13 Prove that f is one-to-one iff for any set C and all functions $h : C \to A$ and $k : C \to A$ we have that $f \circ h = f \circ k$ implies $h = k$.

14 Consider a function $f : \mathbb{Z}/\equiv_n \to \mathbb{Z}/\equiv_k$ defined by $f([x]_n) = [mx]_k$ where n, k and m are positive integers. Show that f is well-defined iff $k \mid mn$.

4.3 BASIC NUMBER THEORY

Many areas of abstract mathematics make use of number theory implicitly. In this section we collect together concepts and results which we have already seen, and will add to it additional ideas from number theory. These concepts include the Division Algorithm, the greatest common divisor and results about prime numbers. We will provide proofs for results that have not yet been seen.

    We have already been introduced to the notion of one integer dividing another and this relation on integers is both reflexive and transitive. Furthermore, if we restrict ourselves to positive integers, then *divides* is also anti-symmetric, and hence is a partial ordering on positive integers. Here are two further properties of *divides*.

Lemma 4.3 *Let $m, n, d \in \mathbb{Z}$.*

    1. If $m \mid n$ and $n \mid m$, then $m = \pm n$.

    2. If $d \mid m$ and $d \mid n$, then $d \mid (am + bn)$ for all $a, b \in \mathbb{Z}$.

Theorem 4.4 (Division Algorithm) *Let $n, d \in \mathbb{Z}$ with $d > 0$. There exist unique $q, r \in \mathbb{Z}$ having the property that $n = dq + r$ with $0 \le r < d$.*

The integer multiples of $n \in \mathbb{Z}$ will be denoted by $n\mathbb{Z}$. For instance,

$$3\mathbb{Z} = \{0, \pm 3, \pm 6, \pm 9, \ldots\}.$$

In the next lemma we summarize some useful facts about $n\mathbb{Z}$.

Lemma 4.4 *Consider the set $n\mathbb{Z}$ for some $n \in \mathbb{N}$.*

1. *If $x, y \in n\mathbb{Z}$, then so are $x + y, x - y \in n\mathbb{Z}$. We say that $n\mathbb{Z}$ is closed under addition and subtraction.*

2. *If X is any non-empty subset of the integers closed under addition and subtraction (see part 1), then $X = n\mathbb{Z}$ for some $n \in \mathbb{N}$.*

Definition 4.12 *Let a and b be two non-zero integers. The integer d is the **greatest common divisor** of a and b, written $d = gcd(a, b)$ if*

1. *$d > 0$*

2. *$d|a$ and $d|b$ (common divisor) and*

3. *$e|a$ and $e|b$ implies $e|d$ (greatest).*

Theorem 4.5 *For any non-zero integers a and b the greatest common divisor of a and b exists and is unique.*

Corollary 4.1 *Let a and b be non-zero integers.*

1. *If $d = gcd(a, b)$, then there exist $x_0, y_0 \in \mathbb{Z}$ such that $d = ax_0 + by_0$.*

2. *$gcd(a, b) = 1$ iff there exist $x_0, y_0 \in \mathbb{Z}$ such that $ax_0 + by_0 = 1$.*

Definition 4.13 *A positive integer p is **prime** if p has exactly two distinct positive divisors, i.e. $p \neq 1$ and if $a > 0$ and $a|p$, then either $a = 1$ or $a = p$.*

Lemma 4.5 *An integer $p > 1$ is prime iff whenever $a, b \in \mathbb{Z}$ and $p|ab$, then either $p|a$ or $p|b$.*

Proof 4.8 *First assume that p is prime and that $p|ab$. Now either $p|a$ or it does not. When $p \nmid a$ we will show $p|b$, thus completing the proof in one direction. Since $p \nmid a$ this implies $gcd(p, a) = 1$ (convince yourself of this). By Corollary 4.1, there exist $x_0, y_0 \in \mathbb{Z}$ such that $px_0 + ay_0 = 1$. Then*

$$b = b(px_0 + ay_0) = p(bx_0) + ab(y_0).$$

*Since $p|p$ and $p|ab$ we have $p|[p(bx_0) + ab(y_0)]$, i.e. $p|b$. For the converse, assume whenever $a, b \in \mathbb{Z}$ and $p|ab$, then either $p|a$ or $p|b$ and assume to the contrary that p is **not** prime. This implies $p = ab$ for some $1 < a, b < p$. Since $p|p$ we have $p|ab$ and so by assumption $p|a$ or $p|b$ which is not possible, since $1 < a, b < p$, thus a contradiction.* □

Corollary 4.2 *If p is prime and $p|(a_1 a_2 \cdots a_n)$, then $p|a_i$ for some i,*
$1 \le i \le n$.

Proof 4.9 *This follows immediately by induction.*

Theorem 4.6 (Fundamental Theorem of Arithmetic) *Any inte-*
ger $a > 1$ is either prime or can be factored uniquely as a product of
primes. More precisely, there exists unique primes $p_1 < p_2 < \cdots < p_n$
and positive integers $e_1, e_2, \ldots, e_n$ such that $a = p_1^{e_1} p_2^{e_2} \cdots p_n^{e_n}$.

EXERCISES

1 Define a relation $\sim$ on $\mathbb{Z}^{>1}$ (integers greater than 1) as follows:
   $n \sim m$ iff There exists a prime p such that $p|m$ and $p|n$. Decide
   whether or not it is an equivalence relation. If it's not, then provide a counterexample. If it is, verify the axioms of an equivalence
   relation and then describe what the equivalence classes represent.

2 Prove Lemma 4.3.2.

3 Decide whether each of the following statements are true. If so,
   give a proof; otherwise give a counter-example.

   a. If $a, b, c \in \mathbb{Z}$ and $a|b$, then $a|(bc)$.

   b. If $a, b \in \mathbb{Z}$ and $a|(b-1)$, then $a|(b^2 - 1)$.

4 Prove Lemma 4.4.1.

5 For an integer a, prove if p is prime and $p|a$, then $gcd(p, a) = p$.

6 For an integer a, prove if p is prime and $p \nmid a$, then $gcd(p, a) = 1$.

7 For an integers a, b, c, prove if $a|c$, $b|c$ and $gcd(a, b) = 1$, then
   $(ab)|c$.

8 For an integers a, b, c, prove $gcd(a, bc) = 1$ iff $gcd(a, b) = 1$ and
   $gcd(a, c) = 1$.

9 For an integers a, b, c, prove if $c|ab$ and $gcd(a, c) = 1$, then $c|b$.

10 Prove that if $d = gcd(n, m)$, then we can express $n = dx$ and
    $m = dy$ for some integers x and y with $gcd(x, y) = 1$.

11 Prove by induction Corollary 4.2.

4.4   MODULO ARITHMETIC

Our focus in this section is the equivalence relation congruence modulo n on $\mathbb{Z}$. First we prove some basic properties of congruence modulo n. This first property of congruence modulo n is in fact characteristic of a more general concept called a **congruence relation**.

Lemma 4.6 *Let* $a, b, c, d \in \mathbb{Z}$. *If* $a \equiv_n c$ *and* $b \equiv_n d$, *then* $a + b \equiv_n c + d$ *and* $ab \equiv_n cd$.

Proof 4.10 *Since* $a \equiv_n c$ *and* $b \equiv_n d$ *we have* $n | (a - c)$ *and* $n | (b - d)$. *Then* $n | [(a - c) + (b - d)]$ *or equivalently* $n | [(a + b) - (c + d)]$ *which implies* $a + b \equiv_n c + d$. *Furthermore*, $n | [(a - c)b + (b - d)c]$ *or equivalently* $n | (ab - cd)$ *which implies* $ab \equiv_n cd$.

Lemma 4.7 *Let* $n > 1$ *and* a *be integers. There exists* $b \in \mathbb{Z}$ *such that* $ab \equiv_n 1$ *iff* $gcd(a, n) = 1$.

Proof 4.11 *Note that* $ab \equiv_n 1$ *iff* $n | (ab - 1)$ *iff* $ab - 1 = nk$ *for some* $k \in \mathbb{Z}$ *iff* $ab + n(-k) = 1$ *for some* $k \in \mathbb{Z}$ *which implies* $gcd(a, n) = 1$, *by Corollary 4.1.2.*
   Assuming $gcd(a, n) = 1$, *by Corollary 4.1.2, there exist* $x_0, y_0 \in \mathbb{Z}$ *such that* $ax_0 + ny_0 = 1$. *Then as we can see from the previous direction, the* b *we are looking for such that* $ab \equiv_n 1$ *is* x_0.

Now let's focus on the collection of equivalence classes $\mathbb{Z} / \equiv_n$. The first questions that need to be addressed is how many distinct classes are there and is there a nice way to represent them? The next result answers these questions.

Lemma 4.8 *For any positive integer* n, *the quotient class*

$$\mathbb{Z} / \equiv_n \quad \text{equals the set } \{ \, [0], [1], \ldots, [n-1] \, \}.$$

Proof 4.12 *First note that any class is equal to one of* $[0], [1], \ldots, [n-1]$, *since if* $m \in \mathbb{Z}$, *using the Division Algorithm, we can write* $m = qn + r$ *for integers* q *and* r *and* $0 \le r < n$. *Now since* $m - r = qn$ *this implies* $n | (m - r)$ *and so* $m \equiv_n r$ *which implies that* $[m] = [r]$ *where* $0 \le r \le n - 1$. *Finally, note that the* $[0], [1], \ldots, [n-1]$ *are all distinct, since if* $[r] = [s]$ *for* $0 \le r, s \le n - 1$, *then* $r \equiv_n s$ *and so* $n | (r - s)$ *which implies* $r - s = nk$ *for some* $k \in \mathbb{Z}$. *But since* $-n < r - s < n$, *the only way this could be possible is if* $r - s = 0$ *or* $r = s$.

We would now like to take these classes of $\mathbb{Z}/\equiv_n$ and define an addition and multiplication for them. The most natural way to proceed would be to define class addition and multiplication as addition and multiplication of representatives, i.e.

$$[a] + [b] = [a + b] \qquad \text{and} \qquad [a] \cdot [b] = [a \cdot b].$$

The problem is that we have to be sure these two binary operations which are functions from $(\mathbb{Z}/\equiv_n) \times (\mathbb{Z}/\equiv_n)$ to $\mathbb{Z}/\equiv_n$ are well-defined. As it turns out there is nothing to fear as is proved below.

Lemma 4.9 *The operations of addition and multiplication in $\mathbb{Z}/\equiv_n$ defined by $[a] + [b] = [a + b]$ and $[a] \cdot [b] = [a \cdot b]$ are well-defined.*

Proof 4.13 *What we must show is that if $[a] = [c]$ and $[b] = [d]$, then $[a] + [b] = [c] + [d]$ and $[a][b] = [c][d]$. But this follows almost immediately from Lemma 4.6.*

Having the peace of mind that these two operations are well-defined, we now list some properties these operations enjoy. These properties describe an important structure in the mathematical field of abstract algebra called a **commutative ring with unity**. We leave the details as a simple exercise which relies heavily on properties of the integers.

Lemma 4.10 *If $a, b, c \in \mathbb{Z}$, then*

1. $[a] + [b] = [b] + [a]$ *and* $[a][b] = [b][a]$.

2. $[a] + ([b] + [c]) = ([a] + [b]) + [c]$ *and* $[a]([b][c]) = ([a][b])[c]$.

3. $[a] + [0] = [a]$ *and* $[a][1] = [a]$.

4. $[a] + [-a] = [0]$.

5. $[a]([b] + [c]) = [a][b] + [a][c]$.

We point out that not every class $[a] \in \mathbb{Z}/\equiv_n$ has a class $[b]$ such that $[a][b] = [1]$. In fact, we see from Lemma 4.7 that such a class $[b]$ exists iff $gcd(a, n) = 1$. A class $[a]$ which has such a $[b]$ is called a **unit**. Some other classes $[a]$ have the property that there is a class $[b]$ such that $[a][b] = [0]$. A class $[a] \neq [0]$ which has such a $[b]$ is called a **zero divisor**.

Example 4.17 *In* $\mathbb{Z}/\equiv_{10}$ *the class* $[3]$ *is a unit since* $[3][7] = [1]$ *and* $[2]$ *is a zero-divisor since* $[2][5] = [0]$.

In fact, the following is true:

Lemma 4.11 *Every class* $[a] \in \mathbb{Z}/\equiv_n$ *not equal to* $[0]$ *is either a unit or a zero divisor.*

Proof 4.14 *Let* $[a] \neq [0]$. *Either* $\gcd(a, n) = 1$ *or* $\gcd(a, n) > 1$. *In the former case, by Lemma 4.7, we know then that* $[a]$ *is a unit. In the latter case, set* $d = \gcd(a, n) > 1$. *Since* $d|a$ *and* $d|n$ *we may write* $a = dk$ *and* $n = dl$ *for some integers* $k, l \in \mathbb{Z}$ *(note that* $l \neq 0$ *since* $n \neq 0$*). Notice that* $[a][l] = [al] = [dkl] = [nk] = [0]$, *since* $nk \equiv_n 0$. *Hence, we see that* $[a]$ *is a zero divisor.*

Note that a Corollary to this result is the fact that every non-zero class in $\mathbb{Z}/\equiv_p$ (where p is prime) is a unit. We now prove a well-known result in number theory proved by the mathematician Pierre de Fermat.

Lemma 4.12 *[Fermat's Little Theorem] Let* $a, p \in \mathbb{Z}$ *with* p *a prime number and* $p \nmid a$. *In* $\mathbb{Z}/\equiv_p$ *we have* $[a]^{p-1} = [1]$, *or equivalently*

$$a^{p-1} \equiv 1 \ (mod \ p).$$

Proof 4.15 *Since* $[a] \neq [0]$ *we know* $[a]$ *is a unit and so there is a class* $[b]$ *with* $[a][b] = [1]$. *Consider the following list of classes:* $[1][a], [2][a], \ldots, [p-1][a]$. *First note that all the classes in this list are distinct, for if* $[r][a] = [s][a]$ *then by multiplying on the right by* $[b]$ *we have* $[r] = [s]$. *Furthermore, these classes are all not equal to* $[0]$, *since if* $[r][a] = [0]$ *then again by multiplying on the right by* $[b]$ *we have* $[r] = [0]$ *which is not the case. Therefore, this list is simply a reordering of the classes* $[1], [2], \ldots, [p-1]$ *and so* $[1][a][2][a] \cdots [p-1][a] = [1][2] \cdots [p-1]$. *This can be rewritten as* $[a]^{p-1}[2] \cdots [p-1] = [2] \cdots [p-1]$. *Since* $[2], \ldots, [p-1]$ *are units, we can reduce this equation to* $[a]^{p-1} = [1]$.

We also have the following result:

Corollary 4.3 *Let* $a, p \in \mathbb{Z}$ *with* p *a prime number. In* $\mathbb{Z}/\equiv_p$ *we have* $[a]^p = [p]$, *or equivalently*

$$a^p \equiv a \ (mod \ p).$$

An equivalent way to represent $\mathbb{Z}/\equiv_n$ with its two operations of addition and multiplication is to consider the following structure. Let $\mathbb{Z}_n = \{0, 1, 2, \ldots, n-1\}$ and define addition, denoted by $+_n$, and multiplication, denoted by $\cdot_n$, for these elements as follows: $m +_n k = r$ if using the Division Algorithm, $m + k = qn + r$ where $0 \leq r < n$. Similarly, $m \cdot_n k = r$ if using the Division Algorithm, $m \cdot k = qn + r$ where $0 \leq r < n$. There is a direct correspondence between $\mathbb{Z}/\equiv_n$ and $\mathbb{Z}_n$. In some sense, which we do not go into here, they are equal. First off a class $[a] \in \mathbb{Z}/\equiv_n$ can be equated with $a \in \mathbb{Z}_n$ by Lemma 4.8. Furthermore, for $0 \leq a, b, c \leq n-1$, $[a] + [b] = [c]$ iff $a +_n b = c$ and $[a][b] = [c]$ iff $a \cdot_n b = c$. We prove this fact for addition (multiplication is similar). Suppose first that $[a] + [b] = [c]$. This implies that $a + b \equiv_n c$ and so $n|(a + b - c)$ which implies $a + b - c = nq$ or $a + b = nq + c$ where $0 \leq c < n$. Then by definition $a +_n b = c$. Now assume that $a +_n b = c$. By definition this implies $a + b = qn + c$ where $0 \leq c < n$. Thus, $qn = a + b - c$ and so $n|(a+b-c)$ which implies $a+b \equiv_n c$ which is equivalent to $[a] + [b] = [c]$.

As a result of this observation, all the properties we proved in this section about $\mathbb{Z}/\equiv_n$ are equally true for $\mathbb{Z}_n$.

EXERCISES

1 Prove Lemma 4.10.

2 Prove that $[a] \in \mathbb{Z}/\equiv_n$ has a class $[b]$ such that $[a][b] = [1]$ iff $gcd(a, n) = 1$.

3 Consider $\mathbb{Z}/\equiv_6$

    a. Write out the addition and multiplication tables

    b. List the units and zero divisors

4 List separately the units and zero divisors of $\mathbb{Z}/\equiv_{12}$. Illustrate this explicitly for each class as we did in Example 4.17.

5 Use Fermat's Little Theorem to compute $[4^{195}]_{13}$.

6 Prove that every non-zero class in $\mathbb{Z}/\equiv_p$ (where p is prime) is a unit.

7 Prove Corollary 4.3.

8 Prove that $[a][b] = [c]$ iff $a \cdot_n b = c$.

4.5 SIZES OF INFINITY

In this section we delve deep into the notion of infinity. In the past this idea of the existence of the infinite had not only mathematical implications but also philosophical and even religious implications. We will show in this section that there are different sizes of infinity, an idea first explored by Georg Cantor. Indeed, when he first proposed this notion that there were different sizes of infinity, mathematicians, philosophers and western religious leaders found it to be anything ranging from nonsense to an abomination.

4.5.1 Countable versus Uncountable

The first distinction Cantor made regarding infinity was that infinite sets can be divided into those which are countable and those which are uncountable. Let's define this terminology.

Definition 4.14 *Let A be a set.*

1. *The set A is **denumerable** if it can be put in one-to-one correspondence with the natural numbers, i.e. there exists a bijection $f : \mathbb{N} \to A$.*

2. *The set A is **countable** if it is either finite or denumerable.*

3. *The set A is **uncountable** if it is not countable, i.e. it is infinite and yet there is no bijection $f : \mathbb{N} \to A$.*

Example 4.18 *Let's give some simple examples illustrating these definitions. We leave the verification of bijection as an exercise.*

1. *Certainly $\mathbb{N}$ is denumerable via the identity map.*

2. *The even numbers are denumerable via the bijection $f : \mathbb{N} \to A$ defined by $f(n) = 2n$. This example may seem counterintuitive since the even numbers are "half the size" of the natural numbers, however in terms of infinity taking half the size of an infinite set does not change its infinite value. This value, which can be studied in detail in a set theory course, is called the **cardinality** of a set.*

3. *The integers are denumerable via the bijection $f : \mathbb{N} \to A$ defined by*

$$f(n) = \begin{cases} -k & \text{if } n = 2k - 1 \\ k & \text{if } n = 2k \end{cases}$$

4. *The rational numbers are denumerable. Indeed, if we can express any set by a well-defined list (explained below), then it must be countable. Consider the following list containing all the rational numbers in lowest terms with only the numerator taking on negative values:*

$$\frac{0}{1}, \frac{-1}{1}, \frac{1}{1}, \frac{-2}{1}, \frac{-1}{2}, \frac{1}{2}, \frac{2}{1}, \frac{-3}{1}, \frac{-1}{3}, \frac{1}{3}, \frac{3}{1}, \frac{-4}{1}, \frac{-3}{2}, \frac{-2}{3}, \frac{-1}{4}, \frac{1}{4}, \frac{2}{3},$$
$$\frac{3}{2}, \frac{4}{1}, \cdots$$

The logic for how we constructed this list of fractions obeys these two rules (page 11 of [6]).

(a) *Define the **height** of a fraction $\frac{m}{n}$ to equal $|m|+n$. In the list above the fractions are listed in ascending height order.*

(b) *Among those fractions of the same height, we list these fractions with numerators in ascending order.*

Then the bijection we seek is define by $f : \mathbb{N} \to \mathbb{Q}$ by

$$f(0) = \frac{0}{1}, \ f(1) = \frac{-1}{1}, \ f(2) = \frac{1}{1}, \ f(3) = \frac{-2}{1}, \ f(4) = \frac{-1}{2}, \cdots$$

5. *The real numbers are not denumerable, and hence uncountable. This verifies the existence of uncountable sets and establishes different sizes of infinity. Indeed, there are "too many" real numbers to put them in one-to-one correspondence with the natural numbers. We will prove this by showing that any function $f : \mathbb{N} \to \mathbb{R}$ cannot map onto $\mathbb{R}$. This type of proof is called a **diagonal argument** which we will afterwards explain how it gets its name. Our proof is nearly identical to Cantor's original proof published in 1891 [1]. We will represent real numbers in terms of their decimal expansion.*

$$z.d_0 d_1 d_2 \cdots, \ \text{where } n \in \mathbb{Z} \text{ and each } d_i \in \{0, 1, 2, \ldots, 9\}.$$

Let's now list the image of f.

$$f(0) = z_0.d_{00}d_{01}d_{02}d_{03}d_{04}d_{05}\ldots,$$

$$f(1) = z_1.d_{10}d_{11}d_{12}d_{13}d_{14}d_{15}\ldots,$$

$$f(2) = z_2.d_{20}d_{21}d_{22}d_{23}d_{24}d_{25}\ldots,$$

$$f(3) = z_3.d_{30}d_{31}d_{32}d_{33}d_{34}d_{35}\ldots,$$

$$f(4) = z_4.d_{40}d_{41}d_{42}d_{43}d_{44}d_{45}\ldots,$$

$$f(5) = z_5.d_{50}d_{51}d_{52}d_{53}d_{54}d_{55}\ldots,$$

$$\vdots$$

In general, $f(n) = z_n.d_{n0}d_{n1}d_{n2}d_{n3}d_{n4}d_{n5}\ldots$.

We will now create a real number which is not in the image of the map f. The real number will be $r = 0.d_0d_1d_2\cdots$, where for every $i = 0, 1, 2, \ldots$

$$d_i = \begin{cases} 2, & \text{if } d_{ii} \neq 2 \\ 8, & \text{if } d_{ii} = 2 \end{cases}$$

Certainly, this real number does not appear as an output of the function f, since it differs from each output in at least one decimal place, namely d_{ii}, for $f(i)$ and $i = 0, 1, 2, \ldots$. Hence, the real number r we just constructed is not in the image of f, and so f does not map onto $\mathbb{R}$.

*The choice of 2 and 8 are somewhat arbitrary. Indeed, Cantor left them as unknowns in his proof. However, we need to steer clear of choosing either 0 or 9 since, for instance, $0.0\bar{9}$ equals 0.1. It is called a **diagonal argument** because we created our real number by examining the diagonal decimal entries of this list of decimal expansions.*

Here are some general facts about countable sets.

Theorem 4.7 *Let A and B be countable sets.*

1. *If there exists a bijection $f : A \to C$ or a bijection $f : C \to A$ for some set C, then C is also countable.*

2. *Any subset of A is countable.*

3. *$A \cup B$ is countable, and thus any countable union of countable sets is countable.*

4. *$A \times B$ is countable, and thus any countable Cartesian product of countable sets is countable.*

5. *Every infinite set has a denumerable subset.*

Proof 4.16 *We leave the first and second statements as an exercises.*

For the third statement, WLOG we may assume A and B are disjoint, since otherwise replace B by $A - B$ which still yields the same union. By assumption there exists bijections $f : \mathbb{N} \to A$ and $g : \mathbb{N} \to B$. Define the map

$$h : \mathbb{N} \to A \cup B \quad defined\ by\ \ h(n) = \begin{cases} f(k), & if\ n = 2k + 1 \\ g(k), & if\ n = 2k \end{cases}$$

It's easy to see that h is in turn a bijection, thus demonstrating that the union is countable.

For the fourth statement, again by assumption there exist bijections from $\mathbb{N}$ to A and from $\mathbb{N}$ to B. By Theorem 4.3.3, the inverses of these two bijections are also bijections. Let's name them $f : A \to \mathbb{N}$ and $g : B \to \mathbb{N}$, and define the map

$$h : A \times B \to \mathbb{N} \times \mathbb{N} \quad defined\ by\ \ h(a,b) = (f(a), f(b)).$$

We leave it as an exercise to show that this function h is also a bijection. Now we define a map

$$k : \mathbb{N} \times \mathbb{N} \to \mathbb{N} \quad by\ \ k(m,n) = \frac{1}{2}\left[(m+n)^2 + 3m + n\right].$$

Let's explain how we came up with this formula. In Figure 4.11, we systematically label all the elements of $\mathbb{N} \times \mathbb{N}$.

$$(0,0) \mapsto 0, \ (0,1) \mapsto 1, \ (1,0) \mapsto 2, (0,2) \mapsto 3, \ (1,1) \mapsto 4,$$

$$(2,0) \mapsto 5, \ (0,3) \mapsto 6, \ (1,3) \mapsto 7, (2,1) \mapsto 8, \ (3,0) \mapsto 9, \ etc..$$

This enumeration of $\mathbb{N} \times \mathbb{N}$ in and of itself defines a bijection from $\mathbb{N} \times \mathbb{N}$ to $\mathbb{N}$, but let's be more explicit about how this function is defined. This enumeration will be our function k. Notice that all the ordered pairs

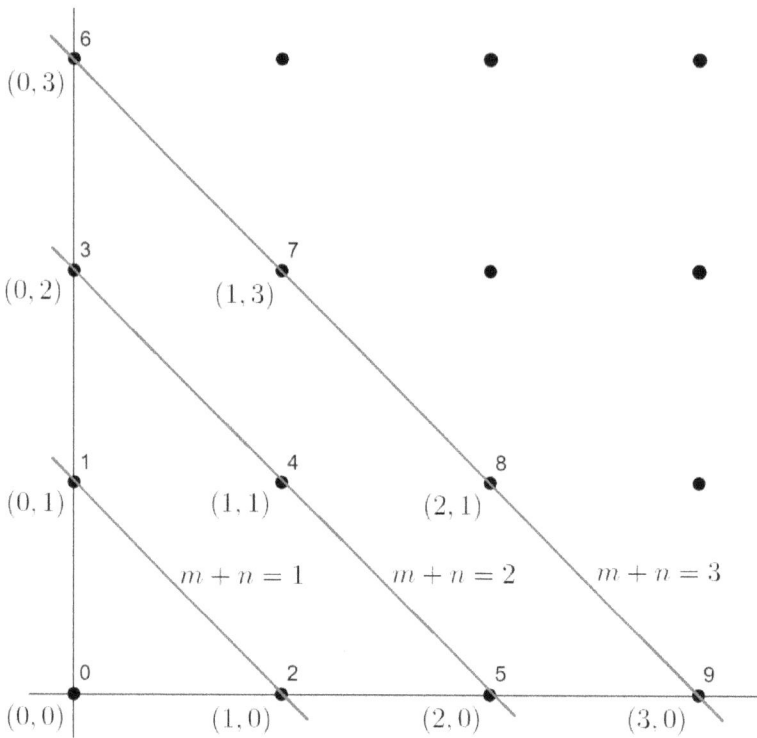

Figure 4.11 Enumerating $\mathbb{N} \times \mathbb{N}$ by $\mathbb{N}$

(m, n) *on a given diagonal line have a fixed* $m+n$ *value, and to count all the points on these diagonal lines (including the origin) up to a certain fixed* $m+n$ *value, we simply sum the numbers from* 1 *up to* $m+n$. *Using Example 2.9,*

$$1 + 2 + \cdots + (m+n) = \frac{1}{2}(m+n+1)(m+n) = \frac{1}{2}\left[(m+n)^2 + m + n\right].$$

If we add additional points on the next diagonal line we must add m, *so that the sum is now*

$$\frac{1}{2}\left[(m+n)^2 + m + n\right] + m = \frac{1}{2}\left[(m+n)^2 + 3m + n\right].$$

This is exactly the formula which defines the one-to-one correspondence, i.e. if we start at 0 *and count the number of ordered pairs diagonal line-by-diagonal line*

$$(m, n) \mapsto \frac{1}{2}\left[(m+n)^2 + 3m + n\right].$$

We leave it as an exercise to verify that k is a bijection. Then by Theorem 4.3.1 and Theorem 4.3.4, the composition $k \circ h : A \times B \to \mathbb{N}$ is a bijection. By the first statement in this theorem we know $A \times B$ countable.

We leave as an exercise the proof of the fifth statement.

□

Remark 4.3 *Here are some additional remarks about countability.*

1. *As we saw in the proof of Theorem 4.7.4, to show a set A is countable we can find a bijection from A to $\mathbb{N}$ as well, which follows from Theorem 4.7.1, since $\mathbb{N}$ is countable.*

2. *Another way to see that $\mathbb{Q}$ is countable is by considering the injection $f : \mathbb{Q} \to \mathbb{Z} \times \mathbb{N}$ by*

$$f \left(\frac{m}{n} \right) = (m, n), \ \ where \ \frac{m}{n} \ is \ in \ lowest \ terms.$$

We leave the verification that f is an injection as an exercise, but note that since $\mathbb{Z}$ and $\mathbb{N}$ are countable, by Theorem 4.7.4, so is $\mathbb{Z} \times \mathbb{N}$. Then the image $f(\mathbb{Q})$ which is a subset of $\mathbb{Z} \times \mathbb{N}$ is therefore countable, by Theorem 4.7.2. Finally, f is a bijection between $\mathbb{Q}$ and its image $f(\mathbb{Q})$, so by Theorem 4.7.1, $\mathbb{Q}$ is countable.

4.5.2 Equinumerous Sets

In the previous subsection we more or less established how to determine if two infinite sets are of the same size when we defined how to decide if a set is the same size as $\mathbb{N}$. We did this by finding a bijection between $\mathbb{N}$ and the set in question. This way of comparing sizes of infinite sets can be generalized. In general (whether finite or infinite) we shall call two sets of the same size **equinumerous**, and we now define this as well as some additional terminology.

Definition 4.15 *Let A an B be two sets.*

1. *A is **equinumerous** to B, written $A \approx B$, if there exists a bijection $f : A \to B$.*

2. *A is **dominated** by B, written $A \preceq B$, if there exists an injection $f : A \to B$.*

3. A is **strictly dominated** *by* B, *written* $A \prec B$, *if* A *is dominated by* B *but not equinumerous to* B, *i.e.* $A \preceq B$ *but* $A \not\approx B$.

Example 4.19 *Let's give some examples illustrating* $\approx$.

1. *Consider the sets* $A = \{1, 2, 3\}$ *and* $B = \{2, 4, 8\}$. *Consider the map* $f : A \to B$ *by* $f(x) = 2^x$. *First,* f *is one-to-one, since if* $f(a) = f(b)$, *i.e.* $2^a = 2^b$, *then* $\log_2(2^a) = \log_2(2^b)$, *i.e.* $a = b$. *Since* $|A| = 3 = |B|$, *by Lemma 4.2.2,* f *is a bijection. Thus,* $A \approx B$.

2. *We've seen in the previous subsection that* $\mathbb{N} \approx \mathbb{Z} \approx \mathbb{Q}$, *since they are all countable.*

3. *The open interval* $(-\pi/2, \pi/2)$ *is equinumerous to* $\mathbb{R}$ *via the map* $f(x) = \tan x$. *Note that* f *is strictly increasing on* $(-\pi/2, \pi/2)$, *since* $f'(x) = \sec^2 x > 0$ *on* $(-\pi/2, \pi/2)$. *Therefore, by the horizontal line test,* f *is one-to-one. Furthermore,* f *maps onto* $\mathbb{R}$, *since*

$$\lim_{x \to -\pi/2} \tan x = -\infty \quad and \quad \lim_{x \to \pi/2} \tan x = \infty.$$

4. *Any open interval* (a, b) *is equinumerous to* $(0, 1)$ *via the map* $f : (0, 1) \to (a, b)$ *by* $f(x) = a + (b - a)x$. *Indeed, the graph of this function is a line segment connecting the point* $(0, a)$ *to the point* $(1, b)$. *We leave it as an exercise to formally show that* f *is a bijection.*

5. *Combining together the results in the previous two examples with the fact the* $\approx$ *is an equivalence relation tells us that any open interval* (a, b) *is equinumerous to* $\mathbb{R}$. *This is somewhat counterintuitive that a bounded subinterval has the same size as the entire real line!*

6. *It is not entirely obvious how to show that, for instance,* $(0, 1) \approx [0, 1]$, *and this will require a very important theorem in order to do so.*

Example 4.20 *Now let's give some examples illustrating* $\prec$ *and* $\preceq$.

1. *Certainly if* $A \subseteq B$ *are sets, then* $A \preceq B$ *via the inclusion map. However, this does not always imply that* $A \approx B$. *Take the example*

of $\mathbb{N} \subseteq \mathbb{R}$ where $\mathbb{N} \preceq \mathbb{R}$ but $\mathbb{N} \not\approx \mathbb{R}$, since $\mathbb{N}$ is countable while $\mathbb{R}$ is uncountable. Indeed, this says that $\mathbb{N} \prec \mathbb{R}$.

2. *One can construct examples where $A \subseteq B$ with $B \preceq A$. Take the example $\mathbb{N} \subseteq \mathbb{Q}^+$, where $\mathbb{Q}^+$ represents the strictly positive rational numbers. Define the map*

$$f : \mathbb{Q}^+ \to \mathbb{N} \quad by \quad f\left(\frac{m}{n}\right) = 2^m 3^n.$$

By the uniqueness of prime factorization, f must be an injection and therefore $\mathbb{Q}^+ \preceq \mathbb{N}$. In fact, from the previous subsection work we know that $\mathbb{Q}^+ \approx \mathbb{N}$, since both are countable.

Remark 4.4 *Here we make some remarks regarding $\approx$, $\preceq$ and $\prec$.*

1. *We make the obvious statement that $\preceq$ is logically equivalent to $\prec \oplus \approx$.*

2. *We leave it as an exercise to show that $\approx$ is an equivalence relation on sets.*

3. *We also leave it as an exercise the fact that $\preceq$ is reflexive, $\prec$ is irreflexive, and both are transitive.*

4. *It's not obvious that $A \preceq B$ and $B \preceq A$ implies $A \approx B$, i.e. that $\preceq$ is anti-symmetric. Indeed, this needs to be proved, and the result is called the Cantor-Shröder-Bernstein Theorem. This we will prove shortly.*

5. *One consequence of the Cantor-Shröder-Bernstein Theorem is that no proper subset of another set can strictly dominate that set. Indeed, if $A \subseteq B$ with $B \prec A$, since $A \preceq B$, by the Cantor-Shröder-Bernstein Theorem, $A \approx B$ which contradicts $B \prec A$.*

Theorem 4.8 (Cantor-Shröder-Bernstein) *For any sets A and B, if $A \preceq B$ and $B \preceq A$, then $A \approx B$.*

Proof 4.17 *Since $A \preceq B$ and $B \preceq A$, there exist injections $f : A \to B$ and $g : B \to A$. This makes $g : B \to g(B)$ a bijection. Therefore, there exists $g^{-1} : g(B) \to B$. Define the following sequence of subsets of A:*

$$A_1 = A - g(B), \quad A_2 = g(f(A_1)), \quad \dots, \quad A_{n+1} = g(f(A_n)) \text{ for } n \geq 1.$$

Now set $X = \cup_{n=1}^{\infty} A_n$ and $Y = A - X$. Certainly X and Y are disjoint. Furthermore, $Y \subseteq g(B)$, since $A_1 = A - g(B)$ and each $A_{n+1} \subseteq g(B)$ for $n \geq 1$. Define $h : A \to B$ by

$$h(a) = \begin{cases} f(a), & \text{if } a \in X \\ g^{-1}(a), & \text{if } a \in Y \end{cases}$$

First, note that h is well-defined, since $X \cap Y = \emptyset$ and $Y \subseteq g(B)$. There are two cases to consider when proving that h is one-to-one (note that this will be a bootstrap cases proof). Suppose that $h(a_1) = h(a_2)$ for some $a_1, a_2 \in A$. WLOG, we can consider just two cases (since a_1 and a_2 are arbitrary).

Case 1: *$a_1 \in X$.*

In this case $a_2 \in X$ as well, for suppose to the contrary that $a_2 \in Y$. This would imply that $f(a_1) = g^{-1}(a_2)$, and applying g to both sides yields $g(f(a_1)) = a_2$. Since $a_1 \in X$, this implies that $a_1 \in A_n$ for some $n \geq 1$. But then $a_2 = g(f(a_1)) \in A_{n+1} \subseteq X$, a contradiction.

Hence, $a_2 \in X$ and we have $f(a_1) = f(a_2)$, and since f is injective, $a_1 = a_2$, which proves the result in this case.

Case 2: *$a_1 \in Y$.*

In this case $a_2 \in Y$ as well, for if $a_2 \in X$, by Case 1, $a_1 \in X$, a contradiction. Therefore, we have $g^{-1}(a_1) = g^{-1}(a_2)$, and since g^{-1} is injective, $a_1 = a_2$, which proves the result in this case.

Now we will show that h maps onto B which will complete the proof. Take any $b \in B$. Again we will consider two cases.

Case 1: *$g(b) \in X$.*

In this case, $g(b) \in A_n$ for some $n \geq 1$. Since $g(b) \in g(B)$ we know that $n > 1$, in which case $g(b) = g(f(a))$ for some $a \in A_{n-1}$. Since g is one-to-one, this implies $b = f(a)$. Now $a \in A_{n-1} \subseteq X$, thus $b = h(a)$ and h maps a onto b, which proves the result in this case.

Case 2: *$g(b) \in Y$.*

In this case $h(g(b)) = g^{-1}(g(b)) = b$ and so h maps $g(b)$ onto b, which proves the result in this case.

□

Example 4.21 *In this example, we give a nice application of the Cantor-Shröder-Bernstein Theorem. We show for any $a, b \in \mathbb{R}$ that*

$$(a, b) \approx [a, b) \approx (a, b] \approx [a, b] \approx \mathbb{R}.$$

In order to do this first we point out that

$$(a, b) \approx (-1, 1), \ [a, b) \approx [-1, 1), \ (a, b] \approx (-1, 1] \ and \ [a, b] \approx [-1, 1].$$

This can be proved by constructing a bijection (similar to Example 4.19.4 which works for all four cases – left as an exercise). Now consider Figure 4.12. Each i_k represents an inclusion map and the map f is defined by

$$f(x) = \frac{x}{|x| + 1}.$$

Now all of these maps are injective (we leave the verification of f as an exercise). Using the Cantor-Shröder-Bernstein Theorem, the sets in the figure are all equinumerous. We will illustrate this with one example, namely $(-1, 1) \approx [-1, 1]$. Now $(-1, 1) \preceq [-1, 1]$ via the injection $i_4 \circ i_1$, and $[-1, 1] \preceq (-1, 1)$ via the injection $f \circ i_3$. Hence, by the Cantor-Shröder-Bernstein Theorem, $(-1, 1) \approx [-1, 1]$.

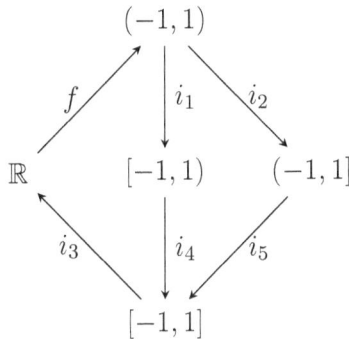

Figure 4.12 A collection of injections between intervals and $\mathbb{R}$.

4.5.3 Infinity Unbounded

We are in a position to show that there are an infinite number of non-equinumerous infinite sets. In fact, one can show that this infinite number of non-equinumerous infinite sets is also uncountable (we will not do that in this text). It all boils down to this simple statement that for any set A, it is always the case that $A \prec \mathcal{P}(A)$, i.e. $A \preceq \mathcal{P}(A)$ but $A \not\approx \mathcal{P}(A)$, where $\mathcal{P}(A)$ is the power set of A – the set of all subsets of A. This is certainly True for finite sets A, since (see Example 2.10)

$$|\mathcal{P}(A)| = 2^{|A|}.$$

Now, if we can prove $A \prec \mathcal{P}(A)$ for infinite sets, then we can create larger and larger sizes of infinity, and therefore achieve our goal.

$$A \prec \mathcal{P}(A) \prec \mathcal{P}(\mathcal{P}(A)) \prec \mathcal{P}(\mathcal{P}(\mathcal{P}(A))) \prec \cdots \text{etc.}$$

Lemma 4.13 *For any set A, $A \prec \mathcal{P}(A)$.*

Proof 4.18 *Certainly, $A \preceq \mathcal{P}(A)$ via the injective map $f : A \to \mathcal{P}(A)$ by $f(a) = \{a\}$. Now we show that no map $g : A \to \mathcal{P}(A)$ can map onto $\mathcal{P}(A)$. In fact, we will construct the element in $\mathcal{P}(A)$ which is not in $g(A)$. This construction is very similar to the construction in Russell's paradox. Set*

$$B = \{a \in A \mid a \notin g(a)\} \in \mathcal{P}(A).$$

Suppose to the contrary that there existed $a_0 \in A$ such that $g(a_0) = B$. There are two possibilities for a_0 both of which lead to a contradiction.

Case 1: $a_0 \in B$.
In this case, by definition of B, $a_0 \notin g(a_0)$, i.e. $a_0 \notin B$, which contradicts the case we are in.

Case 2: $a_0 \notin B$.
In this case, by definition of B, it must be the case that $a_0 \in g(a_0)$, i.e. $a_0 \in B$, which contradicts the case we are in.

□

One final fact we wish to show is that $\mathbb{R} \approx \mathcal{P}(\mathbb{N})$. This also gives a second proof of the fact that $\mathbb{N} \prec \mathbb{R}$ because of Lemma 4.13. First, we need to define some notation.

Definition 4.16 *Given two sets A and B, the set of all functions from A to B will be denoted by B^A. Furthermore, if $B = \{0,1\}$ we write 2^A in place of B^A.*

Lemma 4.14 *$\mathcal{P}(A) \approx 2^A$, for any set A.*

Proof 4.19 *Define the map $g : \mathcal{P}(A) \to 2^A$ as follows: For each $B \in \mathcal{P}(A)$, the function $g(B) : A \to \{0,1\}$ will be defined by*

$$g(B)(x) = \begin{cases} 0, & \text{if } x \notin B \\ 1, & \text{if } x \in B \end{cases}$$

We leave it as an exercise to show that g is a bijection.

□

Theorem 4.9 $\mathbb{R} \approx \mathcal{P}(\mathbb{N})$

Proof 4.20 *We will prove that $\mathbb{R} \preceq \mathcal{P}(\mathbb{N})$ and $\mathcal{P}(\mathbb{N}) \preceq \mathbb{R}$, so that the result then follows from the Cantor-Shröder-Bernstein Theorem.*

First, to show $\mathbb{R} \preceq \mathcal{P}(\mathbb{N})$ we can show $(0,1) \preceq \mathcal{P}(\mathbb{N})$, since $(0,1) \approx \mathbb{R}$. Define a function $f : (0,1) \to \mathcal{P}(\mathbb{N})$ as follows: If $x = 0.d_1 d_2 d_3 \cdots \in (0,1)$, where each $d_i \in \{0,1,2,\ldots,9\}$, then

$$f(x) = \{d_1, d_2 + 10, d_3 + 20, \ldots\}.$$

This function f is certainly one-to-one.

Second, to show $\mathcal{P}(\mathbb{N}) \preceq \mathbb{R}$ we can show $2^{\mathbb{N}} \preceq \mathbb{R}$, by Lemma 4.14. Define a function $g : 2^{\mathbb{N}} \to \mathbb{R}$ as follows: If $f \in 2^{\mathbb{N}}$, i.e. $f : \mathbb{N} \to \{0,1\}$, then

$$g(f) = 0.f(0)f(1)f(2)\cdots.$$

This function g is certainly one-to-one. □

Let's review what we have discovered thus far about the sizes of infinity. We know that

$$\mathbb{N} \prec \mathcal{P}(\mathbb{N}) \approx \mathbb{R} \prec \mathcal{P}(\mathbb{R}) \prec \mathcal{P}(\mathcal{P}(\mathbb{R})) \prec \cdots \text{etc.}$$

The question remains if there are sets of sizes in between this chain of infinite sets. For instance, is there a set A such that

$$\mathbb{N} \prec A \prec \mathbb{R}?$$

This question is called Cantor's Continuum Hypothesis, and Cantor spent a good deal of time attempting to prove that such an A existed, but never succeeded. Another great mathemtician, David Hilbert, claimed to have a proof of the existence of A which later turned out to be a flawed argument. In 1929, mathematician, Kurt Gödel, proved that any axiomatic system of mathematics will have statements which cannot be proven to be True nor proven to be False (called the Incompleteness Theorem). This was an amazing and astonishing fact about mathematics. In 1940, Kurt Gödel, proved that the Continuum Hypothesis could not be proven to be False. Then in 1963, the mathematician Paul Cohen expanding on Gödel's work proved that the Continuum Hypothesis could not be proven to be True. Thus, the Continuum Hypothesis was neither True nor False – one said it was **independence** from the axioms of mathematics (called ZFC – the Zermelo-Fraenkel axioms and the Axiom of Choice).

EXERCISES

1. Verify that the functions defined in Example 4.18 are bijections.

2. Verify that the function defined in Example 4.3.2 is an injection.

3. Prove Theorem 4.7.1.

4. Prove Theorem 4.7.2.

5. Prove the function h in the proof of Theorem 4.7.3 is a bijection.

6. Prove the inductive step if the proof of Theorem 4.7.3.

7. Prove the function h in the proof of Theorem 4.7.4 is a bijection.

8. Prove the function k in the proof of Theorem 4.7.4 is a bijection.

9. Prove the inductive step if the proof of Theorem 4.7.4.

10. Prove Theorem 4.7.5.

11. Prove by induction that any finite union of countable sets is countable.

12. Prove that a countable union of countable sets is countable.

13. Prove by induction that the Cartesian product of a finite number of countable sets is countable.

14. Verify that the map in Remark 4.3.2 is an injection.

15. Verify that $\approx$ is an equivalence relation on sets.

16. Verify that $\prec$ and $\preceq$ are both irreflexive and transitive.

17. Prove that the map defined in Example 4.19.4 is a bijection.

18. Assuming $A \approx B$ for sets A and B,

    a. Show that $A - \{a\} \approx B - \{b\}$, where $a \in A$ and $b \in B$.

    b. Show that $A - C \approx B - D$, if $C \subseteq A$ and $D \subseteq B$ and $C \approx D$.

19 Prove for sets A and B that $A - B \approx B - A$ implies $A \approx B$.

20 Referring to Example 4.21,

   a. Construct the bijection which proves that

   $$(a, b) \approx (-1, 1), \ [a, b) \approx [-1, 1), \ (a, b] \approx (-1, 1] \text{ and}$$
   $$[a, b] \approx [-1, 1].$$

   b. Verify that the map you create is indeed a bijection.

21 Verify that the function f defined in Example 4.21 is an injection. Be sure to check that f maps into the codomain.

22 Verify that the function g in the proof of Lemma 4.14 is a bijection.

4.6 SYMMETRIES AND COMBINATORICS

In this section we dive deeply into the symmetric group with the ultimate goal of achieving an important counting method. We will approach this topic without having to officially introduce a field of mathematics called group theory. This is perhaps the most involved section in the text and thus the reason why it is the very last section.

4.6.1 Permutations

We focus now in more detail on the symmetric group. This group constitutes one of the origins of group theory and is essential, however we prefer not to introduce the general theory of groups, but rather focus only on these permutations. Recall that the symmetric group, S_n, is the collection of bijections of the set $\{1, 2, \ldots, n\}$ onto itself. We now introduce additional notation for representing permutations. The first representation takes the form

$$\sigma = \begin{pmatrix} 1 & 2 & \cdots & n \\ \sigma(1) & \sigma(2) & \cdots & \sigma(n) \end{pmatrix}.$$

The inputs occur on the top row of the notation while the corresponding outputs occur on the bottom row.

Example 4.22 *In S_5, if $\sigma(1) = 2$, $\sigma(2) = 4$, $\sigma(3) = 5$, $\sigma(4) = 1$ and $\sigma(5) = 3$, then*

$$\sigma = \begin{pmatrix} 1 & 2 & 3 & 4 & 5 \\ 2 & 4 & 5 & 1 & 3 \end{pmatrix}.$$

In other words, the top row is the domain and the bottom row is the range. To get σ^{-1} simply flip σ over and reorder the columns according to the top row, i.e.

$$\sigma^{-1} = \begin{pmatrix} 2 & 4 & 5 & 1 & 3 \\ 1 & 2 & 3 & 4 & 5 \end{pmatrix} = \begin{pmatrix} 1 & 2 & 3 & 4 & 5 \\ 4 & 1 & 5 & 2 & 3 \end{pmatrix}.$$

Consider the permutation

$$\tau = \begin{pmatrix} 1 & 2 & 3 & 4 & 5 \\ 2 & 1 & 5 & 4 & 3 \end{pmatrix}.$$

For brevity we will write $\sigma\tau$ for the composition $\sigma \circ \tau$. When computing the composition, in this text, the reader should note that one reads the maps from right to left just as we do with composition of functions. Hence,

$$\sigma\tau = \begin{pmatrix} 1 & 2 & 3 & 4 & 5 \\ 2 & 4 & 5 & 1 & 3 \end{pmatrix}\begin{pmatrix} 1 & 2 & 3 & 4 & 5 \\ 2 & 1 & 5 & 4 & 3 \end{pmatrix} = \begin{pmatrix} 1 & 2 & 3 & 4 & 5 \\ 4 & 2 & 3 & 1 & 5 \end{pmatrix}.$$

The second way to represent permutations is by means of k-cycles. A k-cycle in S_n is a special permutation represented as follows:

$$\sigma = (i_1 \ i_2 \ \cdots \ i_k) \text{ where } \{i_1, i_2, \ldots, i_k\} \subseteq \{1, 2, \ldots, n\}.$$

In this text, k-cycles are read from left to right. Thus, σ is defined as follows:

$$\sigma(i_1) = i_2, \ \sigma(i_2) = i_3, \ \ldots, \ \sigma(i_{k-1}) = i_k, \ \sigma(i_k) = i_1.$$

Note how in the last input-output, $\sigma(i_k) = i_1$, the last element of the cycle gets cycled back to the first element of the cycle (hence, its name). If $m \notin \{i_1, i_2, \cdots, i_k\}$, then it is understood that $\sigma(m) = m$, i.e. any other m is fixed by σ. A 2-cycle is also called a **transposition**.

Example 4.23 *In S_7, if $\sigma = (2\ 4\ 3\ 6\ 1)$, then*

$$\sigma = \begin{pmatrix} 1 & 2 & 3 & 4 & 5 & 6 & 7 \\ 2 & 4 & 6 & 3 & 5 & 1 & 7 \end{pmatrix}.$$

To find the inverse of a k-cycle simply reverse the order of the numbers in the k-cycle. For example, $\sigma^{-1} = (1\ 6\ 3\ 4\ 2)$. Note that a transposition therefore is its own inverse, since

$$(a\ b)^{-1} = (b\ a) = (a\ b).$$

If $\tau = (2\ 5\ 4)(4\ 1\ 5\ 2)(6\ 5\ 1\ 2\ 4)$ a composition of three cycles, then

$$\tau = \begin{pmatrix} 1 & 2 & 3 & 4 & 5 & 6 & 7 \\ 2 & 1 & 3 & 6 & 4 & 5 & 7 \end{pmatrix}.$$

Definition 4.17 *Two cycles are **disjoint** if they have no numbers in common. Formally, $\sigma = (a_1\ a_2\ \cdots\ a_m)$ and $\tau = (b_1\ b_2\ \cdots\ b_n)$ are disjoint if $A \cap B = \emptyset$, where $A = \{a_1,\ a_2,\ \ldots,\ a_m\}$ and $B = \{b_1,\ b_2,\ \ldots,\ b_n\}$.*

Lemma 4.15 *Disjoint cycles commute, i.e. if $\sigma, \tau \in S_n$ are disjoint cycles, then $\sigma\tau = \tau\sigma$.*

Proof 4.21 *Let $\sigma = (a_1\ a_2\ \cdots\ a_m)$ and $\tau = (b_1\ b_2\ \cdots\ b_n)$ be disjoint cycles with sets A and B as in the definition. Our proof is by cases.*

Case 1: *If $c \notin A \cup B$, then*

$$\sigma(\tau(c)) = \sigma(c) = c = \tau(c) = \tau(\sigma(c)).$$

Case 2: *If $a_i \in A$, then $a_i, \sigma(a_i) \notin B$, and thus*

$$\sigma(\tau(a_i)) = \sigma(a_i) = \tau(\sigma(a_i)).$$

Case 3: *If $b_i \in B$, then $b_i, \tau(b_i) \notin A$, and thus*

$$\sigma(\tau(b_i)) = \tau(b_i) = \tau(\sigma(b_i)).$$

Since $\sigma\tau$ and $\tau\sigma$ agree on every input, they are therefore equal. $\quad\square$

Example 4.24 *Consider the following permutation in S_{11}:*

$$\sigma = \begin{pmatrix} 1 & 2 & 3 & 4 & 5 & 6 & 7 & 8 & 9 & 10 & 11 \\ 3 & 1 & 4 & 7 & 8 & 5 & 2 & 6 & 11 & 10 & 9 \end{pmatrix}.$$

We shall illustrate with this example the result we wish to prove next, namely that every permutation can be written as a product of disjoint cycles. We will do this systematically so that a general algorithm is evident and can be used in the proof to follow, i.e. the proof will be constructive.

Start with 1. Notice that σ sends 1 to 3. Then σ sends 3 to 4, 4 to 7, 7 to 2, and σ sends 2 back to 1. So a cycle in σ is $(1\ 3\ 4\ 7\ 2)$. Now pick the smallest number not mentioned in the cycle we constructed, which would be 5. Now σ sends 5 to 8, which is sent to 6, and 6 is sent back to 5. Hence, a second cycle in σ is $(5\ 8\ 6)$. The smallest number not yet mentioned in both cycles constructed is 9 which is sent to 11, which is sent back to 9. Hence, a third cycle in σ is $(9\ 11)$. The last number not yet mentioned is 10 which is sent to itself. This yields a 1-cycle (10). Therefore,

$$(1\ 3\ 4\ 7\ 2)(5\ 8\ 6)(9\ 11)(10).$$

Typically, we drop any 1-cycles from the representation and simply write

$$(1\ 3\ 4\ 7\ 2)(5\ 8\ 6)(9\ 11).$$

Theorem 4.10 *Every permutation can be written as a product of disjoint cycles.*

Proof 4.22 *Let $\sigma \in S_n$. Note that for a positive integer k, σ^k means the k times composition of σ, and if $k = 0$, then σ^0 is the identity map. Consider the following infinite list:*

$$1,\ \sigma(1),\ \sigma^2(1),\ \sigma^3(1),\ \ldots.$$

Since for any k we know $\sigma^k(1) \in \{1, 2, \ldots, n\}$, then there are surely repeats in the above list. Let r be smallest such that $\sigma^r(1)$ is a repeat of an earlier value in the list.

Claim 4.1 *$\sigma^r(1) = 1$.*

By assumption there is a k with $0 \le k < r$ and $\sigma^r(1) = \sigma^k(1)$. Suppose to the contrary that $k > 0$. Then $\sigma^{r-k}(1) = 1$ contradicting *our assumption that $\sigma^r(1)$ is the first repeat. Hence, $k = 0$ and $\sigma^r(1) = \sigma^0(1) = 1$ which proves the claim.*

Therefore, by the claim, one of the cycles in σ's representation as a product of disjoint cycles is

$$(1 \ \sigma(1) \ \sigma^2(1) \ \sigma^3(1) \ \cdots \ \sigma^{r-1}(1)).$$

Certainly, $r \leq n$. If $r = n$ then we are done with the proof with σ being represented by a single cycle. Otherwise, choose the smallest number, say i, not in the list

$$1, \ \sigma(1), \ \sigma^2(1), \ \sigma^3(1), \ \ldots, \ \sigma^{r-1}(1).$$

A similar argument to the one above shows that, for some positive integer s, another one of the cycles in σ's representation as a product of disjoint cycles is

$$(i \ \sigma(i) \ \sigma^2(i) \ \sigma^3(i) \ \cdots \ \sigma^{s-1}(i)).$$

Repeat this process of producing cycles until all of the numbers $1, 2, \ldots, n$ are used up. We leave it as an exercise to verify that we have produced disjoint cycles. □

Remark 4.5 *The representation of a permutation as a product of disjoint cycles is unique up to the order of the cycles (since they commute). Although this statement might be intuitively obvious, a formal proof should really be provided, however we will skip this result.*

Definition 4.18 *The **cycle type** of a given permutation is the length and number of cycles in its unique disjoint cycle representation.*

Example 4.25 *Consider the permutation in the previous example where the disjoint cycle representation was*

$$\sigma = (1\ 3\ 4\ 7\ 2)(5\ 8\ 6)(9\ 11)(10).$$

The cycle type for σ can be expressed as $(*)(**)(* * *)(* * * * *)$. Typically, the cycles are arranged in increasing order of length.

Theorem 4.11 *Every permutation can be written as a product of transpositions.*

Proof 4.23 *To see this result, since we already have Theorem 4.10, it suffices to show that every k-cycle can be written as a product of transpositions. Therefore, consider an arbitrary k-cycle $(a_1 \ a_2 \ \cdots \ a_k)$. Then*

$$(a_1 \ a_2 \ \cdots \ a_k) = (a_{k-1} \ a_k) \cdots (a_2 \ a_k)(a_1 \ a_k).$$

Hence, we expressed any k-cycle as a product of transpositions. □

Example 4.26 *Consider in S_7 the 5-cycle $(2 \ 4 \ 1 \ 6 \ 7)$. Then*

$$(2 \ 4 \ 1 \ 6 \ 7) = (6 \ 7)(1 \ 7)(4 \ 7)(2 \ 7).$$

Now unlike the previous representation as a product of disjoint cycles, the representation of a permutation as a product of transpositions is not unique. Indeed, in the previous example one can add two transpositions of the form $(1 \ 2)$ to the end of the representation, i.e. 001

$$(2 \ 4 \ 1 \ 6 \ 7) = (6 \ 7)(1 \ 7)(4 \ 7)(2 \ 7)(1 \ 2)(1 \ 2).$$

For that matter we could add four such transpositions at the end, or six, or eight, ad infinitum. Although there is no unique representation as a product of transpositions, there is something that remains invariant with respect to the permutation.

Definition 4.19 *A permutation is called **even** if it can be represented as a product of an even number of transpositions. A permutation is **odd** if it can be represented as a product of an odd number of transpositions.*

Theorem 4.12 *A permutation cannot be both even and odd.*

Proof 4.24 *The majority of this result is taken up with the fact that the statement is true for the identity map. We show that the identity map is only even, i.e. if $1 = \tau_1 \tau_2 \cdots \tau_k$ where each τ_i is a transposition, then k is even. This we do now. For each number m which appears in the transpositions consider the following reasoning: Let j be largest such that m appears in τ_j. Note that $j \neq 1$ for otherwise m would not be fixed by the identity map. There are four possibilities for the product $\tau_{j-1}\tau_j$: $(m \ x)(m \ x)$, $(m \ x)(m \ y)$, $(x \ y)(m \ x)$, or $(y \ z)(m \ x)$. For the first possibility, notice that we can simply remove the two transpositions from the product. For the other three properties, we can rewrite the product so* *that m appears first in τ_{j-1} as follows:*

$$(m \ x)(m \ y) = (m \ y)(x \ y),$$

$$(x \ y)(m \ x) = (m \ y)(x \ y),$$

$$(y \ z)(m \ x) = (m \ x)(y \ z).$$

In summary, we either remove two transpositions from the product or we move the last occurrence of m to the left. In the first possibility repeat the process on the next largest occurrence of m (if it exists). In the other three cases repeat the process on $\tau_{j-2}\tau_{j-1}$. Note that for the number m you must eventually be in the first case, for otherwise m would appear first in τ_1 which as we have already pointed out is not possible. Therefore, ultimately we remove m and eventually all transpositions from the identity map in a two-by-two fashion until none are left. Hence, the product must be comprised of an even number of transpositions.

Now suppose that σ is any permutation and $\sigma = \tau_1\tau_2 \cdots \tau_r$ and $\sigma = \tau_1'\tau_2' \cdots \tau_s'$ represented in two ways as a product of transpositions. Equating the two representations we have $\tau_1\tau_2 \cdots \tau_r = \tau_1'\tau_2' \cdots \tau_s'$ and moving them all to one side by multiplying by the inverses of the transpositions we have

$$1 = \tau_r \cdots \tau_2\tau_1\tau_1'\tau_2' \cdots \tau_s'.$$

By the work above we know that $r + s$ must be even, but this implies that either r and s are both even or r and s are both odd. $\qquad \square$

Definition 4.20 *The* **alternating group**, *written $A_n = \{\sigma \in S_n : \sigma \text{ is even}\}$. The set $O_n = \{\sigma \in S_n : \sigma \text{ is odd}\}$.*

Remark 4.6 *We wish to point out several observation regarding A_n.*

1. *The composition of two elements of A_n is an element of A_n, since if $\sigma = \tau_1\tau_2 \cdots \tau_r$ is a product of an even number of transpositions and so is $\tau = \tau_1'\tau_2' \cdots \tau_s'$, then*

$$\sigma\tau = \tau_1\tau_2 \cdots \tau_r\tau_1'\tau_2' \cdots \tau_s',$$

with $r + s$ also even.

2. By Theorem 4.12, we know that S_n is a disjoint union of A_n and O_n.

3. Exactly half of the permutations in S_n $(n \geq 2)$ are even. To see this it is enough to show that $|A_n| = |O_n|$ which we do by defining a bijection between the two sets. Define a function $f : A_n \to O_n$ by $f(\sigma) = \sigma(1\ 2)$. Certainly f maps into O_n since it adds one transposition to the end of any even permutation, thus making it odd. The function is one-to-one, since if $f(\sigma_1) = f(\sigma_2)$, then $\sigma_1(1\ 2) = \sigma_2(1\ 2)$ and by multiplying both sides of the equation by $(1\ 2)$ we have $\sigma_1 = \sigma_2$. Finally, f maps onto O_n, since for any $\tau \in O_n$ notice that $f(\tau(1\ 2)) = \tau(1\ 2)(1\ 2) = \tau$.

Definition 4.21 The **order** of a permutation σ, written $o(\sigma)$ is the smallest positive power k with the property that σ^k is the identity map.

Example 4.27   1. Consider the permutation

$$\sigma = \begin{pmatrix} 1 & 2 & 3 & 4 & 5 \\ 2 & 1 & 4 & 5 & 3 \end{pmatrix}.$$

Check that

$$\sigma^2 = \begin{pmatrix} 1 & 2 & 3 & 4 & 5 \\ 1 & 2 & 5 & 3 & 4 \end{pmatrix},$$

$$\sigma^3 = \begin{pmatrix} 1 & 2 & 3 & 4 & 5 \\ 2 & 1 & 3 & 4 & 5 \end{pmatrix},$$

$$\sigma^4 = \begin{pmatrix} 1 & 2 & 3 & 4 & 5 \\ 1 & 2 & 4 & 5 & 3 \end{pmatrix},$$

$$\sigma^5 = \begin{pmatrix} 1 & 2 & 3 & 4 & 5 \\ 2 & 1 & 4 & 3 & 4 \end{pmatrix},$$

$$\sigma^6 = \begin{pmatrix} 1 & 2 & 3 & 4 & 5 \\ 1 & 2 & 3 & 4 & 5 \end{pmatrix}.$$

Therefore, $o(\sigma) = 6$.

2. *Consider the 4-cycle $\tau = (2\ 1\ 5\ 3)$. Check that*

$$\tau^2 = (1\ 3)(2\ 5), \quad \tau^3 = (1\ 2\ 3\ 5), \quad \tau^4 = (1)(2)(3)(4).$$

Therefore, $o(\tau) = 4$.

We leave as an exercise the following result:

Lemma 4.16 *Consider the symmetric group S_n.*

1. *The order of a k-cycle is k.*

2. *If σ equals a product of disjoint cycles, say $\sigma = \sigma_1\sigma_2\cdots\sigma_m$, then the order of σ is the least common multiple of the orders of σ_1, $\sigma_2,\ldots,\sigma_m$.*

Example 4.28 *With the results in Lemma 4.16 we can compute quickly the order of any permutation.*

1. *Consider the permutation*

$$\sigma = \begin{pmatrix} 1 & 2 & 3 & 4 & 5 & 6 & 7 & 8 & 9 \\ 2 & 5 & 4 & 3 & 8 & 7 & 9 & 1 & 6 \end{pmatrix} = (1\ 2\ 5\ 8)(3\ 4)(6\ 7\ 9).$$

Therefore, $o(\sigma) = lcm(4,2,3) = 12$.

2. *Let $\sigma = (1\ 3\ 7)(2\ 7\ 3)(1\ 4) \in S_7$. Now σ is not represented as a product of disjoint cycles, so we cannot yet use the second part of Lemma 4.16. We first will need to rewrite σ as a product of disjoint cycles. One can compute*

$$\sigma = \begin{pmatrix} 1 & 2 & 3 & 4 & 5 & 6 & 7 \\ 4 & 1 & 2 & 3 & 5 & 6 & 7 \end{pmatrix} = (1\ 4\ 3\ 2).$$

Therefore, σ is a 4-cycle and as such its order is 4.

3. *Using cycle types, we are now in a position to compute the orders of all the elements in S_n (for a given n) and the number of elements of each order. Let's do this for S_5:*

Cycle Type	Order	Number
$(*)(*)(*)(*)(*)$	1	1
$(*)(*)(*)(**)$	2	$\frac{5!}{3!\cdot 2} = 10$
$(*)(**)(**)$	2	$\frac{5!}{2!\cdot 2\cdot 2} = 15$
$(*)(*)(* * *)$	3	$\frac{5!}{2!\cdot 3} = 20$
$(**)(* * *)$	6	$\frac{5!}{2\cdot 3} = 20$
$(*)(* * **)$	4	$\frac{5!}{4} = 30$
$(* * * * *)$	5	$\frac{5!}{5} = 24$

Some explanation is required to understand how the complete list of cycle types are determined and how each of the cycle types is counted. To determine the cycle types, simply consider all different ways (up to commutativity) of expressing 5 as a sum of positive integers:

$$1+1+1+1+1,$$

$$1+1+1+2,$$

$$1+2+2,$$

$$1+1+3,$$

$$2+3,$$

$$1+4,$$

$$5$$

*We call these sums the **partitions** of the number 5. There are two things to consider when counting the number of permutations in a cycle type. First, there are k different ways to represent the same k-cycle. To see this, consider the example 3-cycle*

$$(1\ 2\ 3) = (3\ 1\ 2) = (2\ 3\ 1).$$

Second, if the cycle type contains m disjoint k-cycles, then there are m! ways to order them all of which yield the same permutation (since they are disjoint). Now let's count the cycle type $(*)(**)(**)$. There are 5! ways to fill in the asterisks. Each 2-cycle can be represented in two ways, thus we divide by 2 twice. Furthermore, the two 2-cycles can be ordered in 2! different ways, thus we also divide by 2!.

Example 4.29 *We now make the connection between what is called the dihedral group and the symmetric group.*

1. *For n = 3, the dihedral group, denoted by D_3, consists of three rotations and three reflections of an equilateral triangle. Label the three vertices with the numbers 1, 2, and 3.*

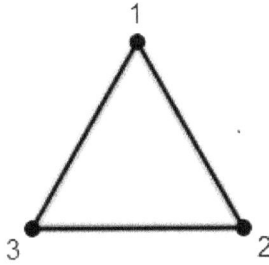

Figure 4.13 Symmetries of the triangle.

The rotations 0°, 120°, and 240° will be denoted respectively by ρ_0, ρ_1 and ρ_2. The reflections will be denoted by μ_1, μ_2 and μ_3 where, for $i = 1, 2, 3$, the reflection μ_i fixes vertex i and swaps the other two vertices. So we can consider the elements of D_3 as permutations of the numbers $1, 2, 3$. Then the elements of D_3 are

$$\rho_0 = \begin{pmatrix} 1 & 2 & 3 \\ 1 & 2 & 3 \end{pmatrix} \qquad \rho_1 = \begin{pmatrix} 1 & 2 & 3 \\ 2 & 3 & 1 \end{pmatrix} \qquad \rho_2 = \begin{pmatrix} 1 & 2 & 3 \\ 3 & 1 & 2 \end{pmatrix}$$

$$\mu_1 = \begin{pmatrix} 1 & 2 & 3 \\ 1 & 3 & 2 \end{pmatrix} \qquad \mu_2 = \begin{pmatrix} 1 & 2 & 3 \\ 3 & 2 & 1 \end{pmatrix} \qquad \mu_3 = \begin{pmatrix} 1 & 2 & 3 \\ 2 & 1 & 3 \end{pmatrix}.$$

Note first that $|D_3| = |S_3|$ and so for $n = 3$ we have $D_3 = S_3$. Let's decide which elements are even and which are odd. We know ρ_0 is even and

$$\rho_1 = (1\ 2\ 3) = (2\ 3)(1\ 3) \qquad \rho_2 = (1\ 3\ 2) = (3\ 2)(1\ 2)$$

$$\mu_1 = (2\ 3) \qquad \mu_2 = (1\ 3) \qquad \mu_3 = (1\ 2).$$

Thus, we see that the rotations are all even and the reflections are all odd and so $A_3 = \{\rho_0,\ \rho_1,\ \rho_2\}$.

2. *For the group D_4 we label the vertices of a square with the numbers 1 through 4 (see Figure 4.14).*

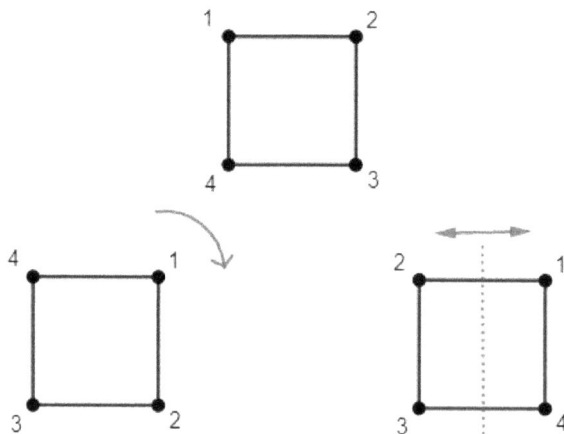

Figure 4.14 Symmetries of the square. On the bottom row, from left to right, are ρ_1 and μ_2

Thus, the elements of D_4 are

$$\rho_0 = \begin{pmatrix} 1 & 2 & 3 & 4 \\ 1 & 2 & 3 & 4 \end{pmatrix} \qquad \rho_1 = \begin{pmatrix} 1 & 2 & 3 & 4 \\ 2 & 3 & 4 & 1 \end{pmatrix}$$

$$\rho_2 = \begin{pmatrix} 1 & 2 & 3 & 4 \\ 3 & 4 & 1 & 2 \end{pmatrix} \qquad \rho_3 = \begin{pmatrix} 1 & 2 & 3 & 4 \\ 4 & 1 & 2 & 3 \end{pmatrix}$$

$$\mu_1 = \begin{pmatrix} 1 & 2 & 3 & 4 \\ 1 & 4 & 3 & 2 \end{pmatrix} \qquad \mu_2 = \begin{pmatrix} 1 & 2 & 3 & 4 \\ 2 & 1 & 4 & 3 \end{pmatrix}$$

$$\mu_3 = \begin{pmatrix} 1 & 2 & 3 & 4 \\ 3 & 2 & 1 & 4 \end{pmatrix} \qquad \mu_4 = \begin{pmatrix} 1 & 2 & 3 & 4 \\ 4 & 3 & 2 & 1 \end{pmatrix}.$$

Note that for $n = 4$ the subscripts of the μ_i do not have the nice relationship to the vertices as they did for $n = 3$. Furthermore, the reader can check that half of the rotations are even and half are odd. The same holds true for the reflections. Since $|D_4| = 8$ and $|S_4| = 4! = 24$, D_4 is properly contained in S_4 and since $|A_4| = 12$ and D_4 has only four even permutations, these four form a proper subset of A_4.

3. *For any $n \geq 3$ there are several general statements we can make. First of all since $|D_n| = 2n$ and $|S_n| = n!$ it is only for $n = 3$ that the two sets coincide. Otherwise D_n is properly contained in S_n and the even permutations in D_n form a proper subset of A_n. Secondly, It is always the case that the rotations are generated by ρ_1, i.e. for all k, $\rho_k = \rho_1^k$.*

EXERCISES

1 Explain why the size of S_n is $n!$.

2 Consider the following permutations in S_6:

$$\sigma = \begin{pmatrix} 1 & 2 & 3 & 4 & 5 & 6 \\ 6 & 2 & 4 & 5 & 1 & 3 \end{pmatrix} \quad \text{and} \quad \tau = \begin{pmatrix} 1 & 2 & 3 & 4 & 5 & 6 \\ 2 & 1 & 4 & 3 & 6 & 5 \end{pmatrix}.$$

   a. Compute $\sigma\tau$, $\tau\sigma$, σ^{-1}, and τ^{-1}.

   b. Express each of σ and τ as a product of disjoint cycles.

   c. Express each of σ and τ as a product of transpositions.

   d. Decide if each of σ and τ is even or odd.

3 Consider the permutation group S_{10} and the following two elements:

$$\sigma = \begin{pmatrix} 1 & 2 & 3 & 4 & 5 & 6 & 7 & 8 & 9 & 10 \\ 4 & 2 & 5 & 8 & 7 & 10 & 3 & 9 & 1 & 6 \end{pmatrix},$$

$$\tau = (1\ 2\ 4)(2\ 5\ 6)(2\ 6\ 8)$$

   a. Write σ as a product of disjoint cycles.

   b. Use part a to compute the order of σ.

   c. Decide whether σ is even or odd.

   d. Compute $\sigma\tau$ and write your answer in the same form as σ was given.

4 Verify that the cycles produced in Theorem 4.10 are disjoint.

5 Prove Lemma 4.16.

6 Apply Lemma 4.16 to find the order of σ in Example 4.24.

7 In S_{11}, apply Lemma 4.16 to find the order of $(1\,3\,2\,6\,8\,11)(5\,7\,9\,10)$

8 Decide which of the elements in D_4 are even and which are odd.

9 For $\sigma, \tau \in S_n$, prove the following statements:

   a. $\tau(a_1\,a_2\,\cdots\,a_k)\tau^{-1} = (\tau(a_1)\,\tau(a_2)\,\cdots\,\tau(a_k))$.

   b. σ and $\tau\sigma\tau^{-1}$ have the same cycle type.

4.6.2 Action

In this subsection we present a way in which the symmetric group S_n may interact with a set of objects. This notion will lead us to a deep result in combinatorics which will allow us to count distinct objects up to a certain collection of permutations. We will explain what we mean by this as we develop the theory and techniques.

Definition 4.22 *We say S_n **acts on** a set X if there is a binary operation $\cdot$ from $S_n \times X$ to X having the following properties:*

   1. For all $\sigma, \tau \in S_n$ and all $x \in X$ we have $\sigma \cdot (\tau \cdot x) = (\sigma\tau) \cdot x$.

   2. For all $x \in X$ we have $1 \cdot x = x$, where 1 represents the identity map in S_n.

*One also says that S_n defines an **action** on X or that X is a S_n-**set**.*

Example 4.30 *Here we present several examples of an action.*

   1. Let $X = \{1, 2, 3, \ldots, n\}$. We can let S_n act on X as follows: for $\sigma \in S_n$ and $i \in X$, define the group action $\sigma \cdot i = \sigma(i)$. For instance, in S_4, if σ is a $90°$ rotation of a square, then $\sigma \cdot 2 = \sigma(2) = 3$. One needs to verify that we have indeed defined a action.

First take $\sigma, \tau \in S_n$ and $i \in X$. Then

$$\sigma \cdot (\tau \cdot i) = \sigma(\tau(i)) = (\sigma\tau)(i) = (\sigma\tau) \cdot i.$$

Second, for the identity permutation 1 and $i \in X$, we have

$$1 \cdot i = 1(i) = i.$$

2. *Let $A = \{a_1, a_2, \ldots, a_n\}$ be any set with n elements. Then for any $k = 1, 2, 3, \ldots$ we can have S_n act on A^k as follows:*

$$\sigma \cdot (a_{i_1}, a_{i_1}, \ldots, a_{i_k}) = (a_{\sigma(i_1)}, a_{\sigma(i_1)}, \ldots, a_{\sigma(i_k)}).$$

Example 4.31 *We now present two important actions which we shall use in later discussions. The reader should verify these examples do indeed define actions.*

1. *Let S_n act on itself (i.e. $X = S_n$) by left multiplication, i.e. for $\sigma \in S_n$ and $\tau \in X$ define $\sigma \cdot \tau = \sigma\tau$, i.e. composition in S_n.*

2. *Let S_n act on itself (i.e. $X = S_n$) by conjugation, i.e. for $\sigma \in S_n$ and $\tau \in X$ define $\sigma \cdot \tau = \sigma\tau\sigma^{-1}$.*

We now define several important structures associated with an action. From now on we will leave out the binary action operator $\cdot$ and simply concatenate a permutaion with a set element to indicate the action. The context will make it clear if we are composing permutations or having a permutation act on a set element.

Definition 4.23 *Let S_n act on a set X with $\sigma \in S_n$ and $x \in X$. For ease of reading, set $G = S_n$.*

1. *The **stabilizer** of x, written $G_x = \{\sigma \in G : \sigma x = x\}$, i.e. the permutations which fix a particular set element x.*

2. *The **fixator** of σ, written $X_\sigma = \{x \in X : \sigma x = x\}$, i.e. the set elements fixed by a particular permutation σ.*

3. *The **orbit** of x, written $Gx = \{\sigma x : \sigma \in G\}$, i.e. the elements of the set that can be realized by allowing all of the permutations to act on a fixed set element x.*

Note that G_x is contained in G while both X_g and Gx are subsets of X.

Example 4.32 *Let us return now to Example 4.31 and compute the structures we just defined.*

1. *G_σ contains only 1, $X_\sigma = \emptyset$ unless $\sigma = 1$ in which case $G_1 = X_1 = G$. It's always the case that $G\sigma = G$.*

2. *$G_\sigma = \{\tau \mid \sigma\tau = \tau\sigma\}$, i.e. all the elements of G which commute with σ, and so is X_g. Now $G\sigma = \{\tau\sigma\tau^{-1} : \tau \in G\}$ is called the* **conjugacy class** *of σ in G. We leave it as an exercise to show that the division of permutations into cycle types corresponds exactly to the conjugacy classes formed by this particular action.*

It is useful to note that orbits can be defined as the equivalence classes of a particular equivalence relation of the set X upon which S_n acts. Define the following relation on X: $x \sim y$ iff there is a $\sigma \in S_n$ such that $\sigma x = y$. This defines an equivalence relation on X with equivalence classes being precisely the orbits of S_n acting on X. Indeed, $\sim$ is reflexive since for any $x \in X$ we have $1x = x$ (definition of an action). We have symmetry, since if $x \sim y$ then there is a $\sigma \in S_n$ with $\sigma x = y$, but then using the definition of an action this can be rewritten as $\sigma^{-1}y = x$ and so $y \sim x$. For transitivity, if $x \sim y$ and $y \sim z$ then there are $\sigma, \tau \in S_n$ with $\sigma x = y$ and $\tau y = z$. But then $(\tau\sigma)x = z$ by using the definition of an action, and thus $x \sim z$. If we take any $x \in X$ and compute the equivalence class

$$[x] = \{y \in X : y \sim x\} = \{y \in X : \exists \sigma \in S_n, y = \sigma x\}$$

$$= \{\sigma x : \sigma \in S_n\} = Gx.$$

One use of this observation is the immediate result that any two orbits of an action are either disjoint or coincide (since equivalence classes have this property).

We now begin our discussion on counting results in this setting. Before we do this, we need to define another structure related to S_n.

Definition 4.24 *Let $G = S_n$ act on a set X with $\sigma \in G$ and $x \in X$. A* **coset** *of G with respect to G_x, written*

$$\sigma G_x = \{\sigma\tau \mid \text{for } \tau \in G_x\}.$$

The collection of all cosets of G with respect to G_x is denoted by G/G_x.

Remark 4.7 *We make several remarks about cosets which we leave as exercises. Let $G = S_n$ acts on a set X.*

1. *Cosets of G with respect to G_x are all the same size, i.e. if $\sigma, \tau \in G$ and $x \in X$, then $|\sigma G_x| = |\tau G_x|$. Since $G_x = 1G_x$, this means all cosets are the same size as G_x (which is how one proves this remark).*

2. *Cosets of G with respect to G_x form equivalence classes of G via the relation $\sigma \sim \tau$ iff $\sigma \tau^{-1} \in G_x$.*

3. *From the previous two remarks, it follows that $|G/G_x| = |G|/|G_x|$. Indeed, set $n = |G/G_x|$. We know that since G is a disjoint union of cosets,*

$$|G| = |\sigma_1 G_x| + |\sigma_2 G_x| + \cdots + |\sigma_n G_x|$$

$$= |G_x| + |G_x| + \cdots + |G_x| = n|G_x| = |G/G_x||G_x|,$$

from which the result follows.

Theorem 4.13 *Let $G = S_n$ act on a set X. If X is finite, then we have $|G| = |G_x||Gx|$. In particular, the size of an orbit divides the size of G.*

Proof 4.25 *We simply define a map $f : G/G_x \to Gx$ by $f(\sigma G_x) = \sigma x$. First, note that f certainly maps onto Gx by its very definition. Second, f is both well-defined (which needs to be checked, since the domain consists of equivalence classes) and one-to-one, since*

$$\sigma G_x = \tau G_x \quad iff \quad \tau^{-1}\sigma \in G_x \quad iff \quad \tau^{-1}\sigma x = x$$

$$iff \quad \sigma x = \tau x \quad iff \quad f(\sigma G_x) = f(\tau G_x).$$

Thus, f is a bijection which proves that $|G/G_x| = |Gx|$. Now use Remark 4.7.3 to complete the proof. $\square$

Example 4.33 *Consider the cycle types (conjugacy classes) of elements in S_4.*

CycleType	Number
$(*)(*)(*)(*)$	1
$(*)(*)(**)$	$\frac{4!}{2!\cdot 2} = 6$
$(**)(**)$	$\frac{4!}{2!\cdot 2 \cdot 2} = 3$
$(*)(***)$	$\frac{4!}{3} = 8$
$(****)$	$\frac{4!}{4} = 6$

Notice how the sizes of the conjugacy classes divide the order of the group S_4 (which equals $4! = 24$), since conjugacy classes are orbits.

EXERCISES

1 Verify that Example 4.30.2 defines an action.

2 Verify that each action defined in Example 4.31 is indeed an action.

3 Verify the statements made in Example 4.32. Note that to show that the division of permutations into cycle types corresponds exactly to the conjugacy classes you will need Exercize 9 of Subsection 4.6.1.

4 Prove the first two remarks made in Remark 4.7.

5 Let $G = S_4$ and set X equal to the set of all transpositions in S_4. Set $x = (2\ 4) \in X$. Let G act on X as follows: For $\sigma \in G$ and $(ij) \in X$ define $\sigma(ij) = (\sigma(i)\sigma(j))$:

   a. Verify that the above definition does indeed define an action.

   b. List the elements of G_x and thus compute $|G_x|$.

   c. List the elements of Gx and thus compute $|Gx|$.

   d. Now compute $|Gx|$ using Theorem 4.13.

6 Consider $G = S_n$ acting on a set X.

    a. For $\sigma \in G$ and $x \in X$ show that $G_{\sigma x} = \sigma^{-1} G_x \sigma$.

    b. For $\sigma \in G$ and $x \in X$ show that $|G_x| = |\sigma^{-1} G_x \sigma|$.

    c. For $x, y \in X$ show that if $Gx = Gy$ then $|G_x| = |G_y|$.

7 S_n acts transitively on a set X if for any $x, y \in X$, there exists a $\sigma \in S_n$ such that $\sigma x = y$. Furthermore, S_n acts doubly transitive on a set X if for every x_1, y_1, x_2, y_2 with $x_1 \neq y_1$ and $x_2 \neq y_2$, there exists a $\sigma \in S_n$ such that $\sigma x_1 = x_2$ and $\sigma y_1 = y_2$.

Show that $G = S_n$ acts doubly transitively on a set X iff for any $x \in X$, G_x acts transitively on $X - \{x\}$ and G acts transitively on X.

4.6.3 Burnside's Lemma

In this subsection we introduce a result, called Burnside's Lemma, which has surprising consequences in combinatorics allowing us to count various things with ease.

Lemma 4.17 (Burnside's Lemma) *If S_n is acting on a finite set X, then the number of orbits of X equals*

$$\frac{1}{|S_n|} \sum_{\sigma \in S_n} |X_\sigma|.$$

Proof 4.26 *Set $G = S_n$. Define a function $f : G \times X \to X$ by*

$$f(\sigma, x) = \begin{cases} 1, & \sigma x = x \\ \\ 0, & \sigma x \neq x \end{cases}$$

Notice that

$$\sum_{\sigma \in G} |X_\sigma| = \sum_{\sigma \in G} \left(\sum_{x \in X} f(\sigma, x) \right) = \sum_{x \in X} \left(\sum_{\sigma \in G} f(\sigma, x) \right)$$

$$= \sum_{x \in X} |G_x| = \sum_{x \in X} \frac{|G|}{|Gx|} = |G| \left(\sum_{x \in X} \frac{1}{|Gx|} \right).$$

Furthermore, consider a typical orbit $Gx = \{x_1, \ldots, x_r\}$ and notice that

$$\sum_{i=1}^{r} \frac{1}{|Gx_i|} = \sum_{i=1}^{r} \frac{1}{|Gx|} = \frac{r}{|Gx|} = 1.$$

Therefore,

$$\sum_{\sigma \in G} |X_\sigma| = |G| \left(\sum_{x \in X} \frac{1}{|Gx|} \right) = |G| \times (\textit{The number of orbits of } X).$$

From this last equation the result follows. □

Remark 4.8 *We wish to point out, without proof (to avoid having to introduce too much mathematical machinery), that Burnside's Lemma holds for any subset of S_n closed under composition, i.e. $H \subseteq S_n$ with the property that $\sigma, \tau \in H$ implies $\sigma\tau \in H$. Two examples of interest are D_n, the collection of rotations and reflections of an n-gon, or just the rotations.*

Example 4.34 *We now give some nice counting arguments which use Burnside's Lemma.*

1. *Consider the letters a, a, b and suppose we wish to count the number of distinct (nonsense) words we can produce using all three letters (these are called **arrangements**). For instance, one such word could be aba. The solution to this problem is easy to count, but we use it to illustrate this counting method. Consider $G = S_3$ acting on the set X of all arrangements of the three letters in the natural way, i.e. for instance, if σ is the 3-cycle $(1\ 2\ 3)$, then $\sigma(aba) = aab$.*

   We will count the number of arrangements of the letters a, a, b. In this case $G = S_3$ which consists of

   $$1, (1\ 2\ 3), (1\ 3\ 2), (2\ 3), (1\ 3), (1\ 2).$$

   One can check that their corresponding fixators are

   $$X_1 = X, \quad X_{(1\ 2\ 3)} = \emptyset, \quad X_{(1\ 3\ 2)} = \emptyset,$$
   $$X_{(2\ 3)} = \{baa\}, \quad X_{(1\ 3)} = \{aba\}, \quad X_{(1\ 2)} = \{aab\}.$$

   Therefore, since there is one orbit,

   $$\frac{1}{|G|} \sum_{g \in G} |X_g| = \frac{1}{3!}(|X| + 0 + 0 + 1 + 1 + 1) = 1.$$

Hence, $|X| = 3$, which is indeed the case, namely the arrangements aab, aba and baa. Granted this is an example you can easily count by hand, but our technique can generalize to any number of letters.

2. *Consider a circle divided into 6 equal sectors as in Figure 4.15. Suppose we can color a sector either black or white and we wish*

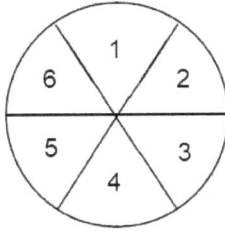

Figure 4.15 Six numbered sectors in a circle.

*to count the number of distinct ways of coloring the circle in the sense that two colorings of the circle are distinct if you cannot get from one to the other by rotating the circle. One says distinct **up to rotation**. To do this we label the sectors with numbers.*

For instance, in Figure 4.16, these two coloring are not distinct, since one can rotate the circle on the left by 120° clockwise and align it with the coloring on the right.

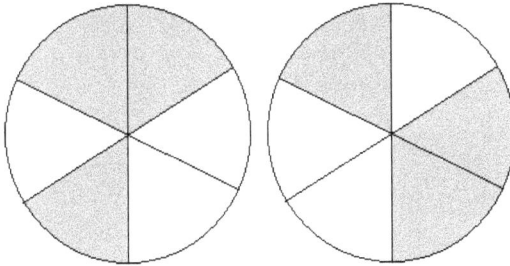

Figure 4.16 Two non-distinct colorings of the circle.

We consider the subset H of S_6 consisting of the six rotations of the circle 0°, 60°, 120°, 180°, 240°, and 300°. These elements of S_6 are

$$\sigma_0 = 1, \qquad \sigma_1 = (1\ 2\ 3\ 4\ 5\ 6), \qquad \sigma_2 = (1\ 3\ 5)(2\ 4\ 6),$$

$$\sigma_3 = (1\ 4)(2\ 5)(3\ 6), \qquad \sigma_4 = (1\ 5\ 3)(2\ 6\ 4), \qquad \sigma_5 = (1\ 6\ 5\ 4\ 3\ 2).$$

Notice again that in Figure 4.16 the permutation taking the coloring of the circle on the left to the coloring on the right is σ_2.

*Let X be the collection of all possible ways to color sectors $1, 2, 3, 4, 5$ and 6. The size of X is $2^6 = 64$, since there are two choices of colors for each of the six sectors in the circle. Let H act on X in the natural way by rotating the circle the appropriate number of degrees, i.e. σ_i will rotate the circle $(60i)°$. Notice that for a given coloring of the circle, any orbit of this coloring is precisely the set of colorings which are **not** considered distinct. Therefore, if we wish to count the number of distinct colorings of the circle, then we need only count the number of orbits of this action. Burnside's Lemma does exactly this thing. Therefore, we first need to compute the six fixators. Certainly, the fixator of identity is the entire set X. The fixator of σ_1 consists of the circle with all white sectors and the circle with all black sectors. The fixator of σ_2 consists of the circle with all white sectors, the circle with all black sectors, the circle with even numbered sectors are one color and the odd numbered sectors the other color. In other words, once you have decided on the colors for sector 1 and 2, then the rest are completely determined. Thus, this gives you $2^2 = 4$ possible colorings in the fixator. The fixator of σ_3 consists of the circles in which opposite sectors are the same color. In other words, once you have decided on the colors for sector 1, 2, and 3, then the rest are completely determined. Thus, this gives you $2^3 = 8$ possible colorings in the fixator. Finally $X_{\sigma_4} = X_{\sigma_2}$ and $X_{\sigma_5} = X_{\sigma_1}$. Therefore, using Burnside's Lemma, the number of orbits equals*

$$\frac{1}{|H|} \left(|X_{\sigma_0}| + |X_{\sigma_1}| + |X_{\sigma_2}| + |X_{\sigma_3}| + |X_{\sigma_4}| + |X_{\sigma_5}| \right)$$

$$= \frac{1}{6}(64 + 2 + 4 + 8 + 4 + 2) = 14.$$

The reader may wish to list the 14 distinct colorings. An approach to generating this list might be to first consider distinct colorings with no black sectors, then one black sector, then two, then three, up to six black sectors.

3. *Consider the same setup as Example 2, but in addition assume that the coloring of a circle shows through to the back of the circle. If we wish to count the number of distinct colorings this means*

we are counting the number of distinct colorings up to rotation and reflection. There are exactly six rigid reflections of the circle, namely the reflections across the three diameters marked on the circle and the reflections across the three diameters which bisect opposite sectors. Name these six reflections to be $\sigma_6, \ldots, \sigma_{11}$ and note that $D_6 = \{\sigma_0, \sigma_1, \ldots, \sigma_{11}\}$. The reader should check that the first three reflections have fixators of size 8 while the second three have fixators of size 16. Therefore, using Burnside's Lemma, the number of orbits equals

$$\frac{1}{|D_6|} \sum_{i=0}^{11} |X_{\sigma_i}| = \frac{1}{12}(64+2+4+8+4+2+8+8+8+16+16+16) = 13.$$

Perhaps surprising, this additional restriction on distinctness barely reduces their number. The reader should look at the list of distinct colorings compiled in Example 2 and decide which two of the 14 colorings are now being equated.

EXERCISES

1 Count the arrangements of the letters a, a, b, c, c, c as in Example 4.34.

2 Repeat Example 4.34.1 and Example 4.34.2, but for the square in Figure 4.17.

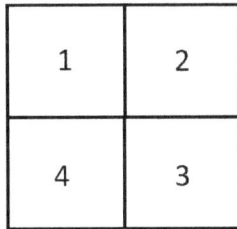

Figure 4.17 Four numbered subsquares in a square.

3 Repeat Example 4.34.1 and Example 4.34.2, but for a circle with five equal sectored number 1 through 5.

4 We wish to paint the roof of a house (see Figure 4.18). There are four sections of roof each of which can be painted in one of two

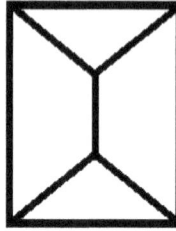

Figure 4.18 The roof of a house.

colors: sienna and ochre. Our goal is to investigate the number of distinct colorings up to (two) rotations and (two) reflections.

Use Burnside's Lemma to count the number of distinct colorings.

5 Count the number of unique dominos first using a simple combinatorial argument, then by using Burnsid'e Lemma.

4.6.4 Polya's Formula

We first define Polya's Formula and then we will apply it to counting distinct colorings. It arises from Burnside's Lemma for counting orbits. Polya's Formula is derived from the cycle types in the permutations acting on the colorings. Each cycle type found corresponds to a monomial in Polya's Formula.

Example 4.35 *If a permutation in S_{12} has cycle type*

$$(*)(*)(**)(**)(**)(****),$$

the corresponding monomial in Polya's Formula is $x_1^2 x_2^3 x_4$.

In general, if a permutation has cycle type which includes exactly n cycles each of length m, then x_m^n is included in the monomial in Polya's Formula corresponding to that cycle type. The coefficient of this monomial will be the number of permutations acting on the colorings of that particular cycle type.

Definition 4.25 *Let H be a subset of S_n closed under composition and acting on a set of colorings. Polya's Formula is a polynomial in unknowns $x_1, x_2, \ldots, x_n$ of the form*

$$\frac{1}{|H|} \sum_{\sigma \in G} x_1^{e_1} x_2^{e_2} \cdots x_n^{e_n},$$

where in the formula above, the σ has e_i cycles of length i in its cycle type (for $i = 1, 2, \ldots, n$).

Example 4.36 *Consider the example from the previous section for coloring the circle with six sectors.*

Permutation	Cyclic Decomposition	Cycle Type	Polya Monomial
$0°$	$(1)(2)(3)(4)(5)(6)$	$(*)(*)(*)(*)(*)(*)$	x_1^6
$60°$	$(1\ 2\ 3\ 4\ 5\ 6)$	$(*****\,*)$	x_6
$120°$	$(1\ 3\ 5)(2\ 4\ 6)$	$(*\,*\,*)(*\,*\,*)$	x_3^2
$180°$	$(1\ 4)(2\ 5)(3\ 6)$	$(**)(**)(**)$	x_2^3
$240°$	$(1\ 5\ 3)(2\ 6\ 4)$	$(*\,*\,*)(*\,*\,*)$	x_3^2
$300°$	$(1\ 6\ 5\ 4\ 3\ 2)$	$(*****\,*)$	x_6

Therefore, Polya's Formula is

$$P(x_1, x_2, x_3, x_6) = \frac{1}{6}(x_1^6 + 2x_6 + 2x_3^2 + x_2^3)$$

Example 4.37 *Considering the same set up as in Example 4.36, if our group is the six rotations and six reflections, then we add the following six rows to the table:*

Permutation	Cyclic Decomposition	Cycle Type	Polya Monomial
μ_1	$(1)(4)(2\ 6)(3\ 5)$	$(*)(*)(**)(**)$	$x_1^2 x_2^2$
μ_2	$(2)(5)(1\ 3)(4\ 6)$	$(*)(*)(**)(**)$	$x_1^2 x_2^2$
μ_3	$(3)(6)(2\ 4)(1\ 5)$	$(*)(*)(**)(**)$	$x_1^2 x_2^2$
μ_4	$(1\ 2)(3\ 6)(4\ 5)$	$(**)(**)(**)$	x_2^3
μ_5	$(2\ 3)(1\ 4)(5\ 6)$	$(**)(**)(**)$	x_2^3
μ_6	$(3\ 4)(2\ 5)(1\ 6)$	$(**)(**)(**)$	x_2^3

Therefore, Polya's Formula is

$$P(x_1, x_2, x_3, x_6) = \frac{1}{12}(x_1^6 + 2x_6 + 2x_3^2 + 4x_2^3 + 3x_1^2x_2^2).$$

As you probably noticed, Polya's Formula looks very similar to Burnside's Lemma for counting orbits. In fact, each monomial corresponds to a fixator. Polya's Formula has two primary uses: To count the number of distinct colorings and to produce the inventory of unique colorings. First, to count the number of distinct colorings simply evaluate all the x_i with the number of colors used. This makes sense since all the numbers in a cycle must be colored with the same color to remain fixed, thus there are as many ways to color the numbers in that cycle as there are colors.

Example 4.38 *Consider again Example 4.36 and Example 4.37.*

1. *If we are looking for the distinct colorings up to rotation, we have seen that Polya's Formula is*

$$P(x_1, x_2, x_3, x_6) = \frac{1}{6}(x_1^6 + 2x_6 + 2x_3^2 + x_2^3).$$

So the number of distinct colorings with two colors is

$$P(2, 2, 2, 2) = \frac{1}{6}(2^6 + 2 \cdot 2 + 2 \cdot 2^2 + 2^3) = 14,$$

which got us to our answer much quicker than Burnside's Lemma. In fact, now we can easily compute distinct colorings with three colors to be

$$P(3, 3, 3, 3) = \frac{1}{6}(3^6 + 2 \cdot 3 + 2 \cdot 3^2 + 3^3) = 130.$$

2. *If we are looking for the distinct colorings up to rotation and reflection, we have seen that Polya's Formula is*

$$P(x_1, x_2, x_3, x_6) = \frac{1}{12}(x_1^6 + 2x_6 + 2x_3^2 + 4x_2^3 + 3x_1^2x_2^2).$$

So the number of distinct colorings with two colors is

$$P(2, 2, 2, 2) = \frac{1}{12}(2^6 + 2 \cdot 2 + 2 \cdot 2^2 + 4 \cdot 2^3 + 3 \cdot 2^2 \cdot 2^2) = 13.$$

Again, we can easily compute distinct colorings with three colors to be

$$P(3, 3, 3, 3) = \frac{1}{12}(3^6 + 2 \cdot 3 + 2 \cdot 3^2 + 4 \cdot 3^3 + 3 \cdot 3^2 \cdot 3^2) = 92.$$

Now let's address the second use of Polya's Formula, namely to produce the inventory of all distinct colorings. Polya's Formula will be used as a *generating function* to list all the possible colorings. How it works is as follows: Let $P(x_1, x_2, \ldots, x_n)$ be Polya's Formula – a multi-variate polynomial in the unknowns $x_1, x_2, \ldots, x_n$. Suppose our colors are $c_1, c_1, \ldots, c_m$. If we evaluate each x_k at $\sum_{i=1}^{m} c_i^k$ we will get a polynomial in $c_1, c_2, \ldots, c_m$ which will describe explicitly the full inventory of distinct colorings. Indeed, replacing x_k by $\sum_{i=1}^{m} c_i^k$ is saying we must color all the elements in a k-cycle the same color, having m colors to choose from. Having done so, the coefficient of $c_1^{e_1} c_2^{e_2} \cdots c_n^{e_n}$ in $P(\sum_{i=1}^{m} c_i, \sum_{i=1}^{m} c_i^2, \ldots, \sum_{i=1}^{m} c_i^n)$ corresponds to the number of ways to color using e_1 colored as c_1, e_2 colored as c_2, ..., e_m colored as c_m.

Example 4.39 *Let's return to Example 4.36 and Example 4.37.*

1. *For distinct colorings of the six sectors up to rotation, we derived Polya's Formula $P(x_1, x_2, x_3, x_6) = \frac{1}{6}(x_1^6 + 2x_6 + 2x_3^2 + x_2^3)$. Suppose first we are coloring using black and white. To get the full inventory of colorings we evaluate*

$$P(b + w, b^2 + w^2, b^3 + w^3, b^6 + w^6)$$

$$= \frac{1}{6}((b + w)^6 + 2(b^6 + w^6) + 2(b^3 + w^3)^2 + (b^2 + w^2)^3)$$

$$= w^6 + bw^5 + 3b^2w^4 + 4b^3w^3 + 3b^4w^2 + b^5w + b^6.$$

What this is telling us is that there is one way to color them all white, one way to color using one black and five white, three ways to color using two black and four white, etc. Notice we still have to decide what the colorings are, but at least we know how many we are looking for of each type. Notice also that if we set $b = 1$ and $w = 1$ in the resulting polynomial, we once again get the number of distinct colorings. Using three colors b, w, r we begin to see the true power of Polya's Formula:

$$P(b + w + r, b^2 + w^2 + r^2, b^3 + w^3 + r^3, b^6 + w^6 + r^6)$$

$$= \frac{1}{6}((b+w+r)^6 + 2(b^6 + w^6 + r^6) + 2(b^3 + w^3 + r^3)^2 + (b^2 + w^2 + r^2)^3)$$

$$= w^6 + rw^5 + bw^5 + 3r^2w^4 + 5brw^4 + 3b^2w^4 + 4r^3w^3 + 10br^2w^3$$

$$+ 10b^2rw^3 + 4b^3w^3 + 3r^4w^2 + 10br^3w^2 + 16b^2r^2w^2 + 10b^3rw^2 + 3b^4w^2$$

$$+ r^5w + 5br^4w + 10b^2r^3w + 10b^3r^2w + 5b^4rw + b^5w + r^6 + br^5 + 3b^2r^4$$

$$+ 4b^3r^3 + 3b^4r^2 + b^5r + b^6.$$

2. *Let's repeat the process for colorings distinct up to rotation and reflection. We derived Polya's Formula in this case to be*

$$P(x_1, x_2, x_3, x_6) = \frac{1}{12}(x_1^6 + 2x_6 + 2x_3^2 + 4x_2^3 + 3x_1^2 x_2^2).$$

Thus, for two colors we evaluate

$$P(b+w, b^2+w^2, b^3+w^3, b^6+w^6)$$

$$= \frac{1}{12}((b+w)^6 + 2(b^6+w^6) + 2(b^3+w^3)^2 + 4(b^2+w^2)^3$$

$$+3(b+w)^2(b^2+w^2)^2)$$

$$= w^6 + bw^5 + 3b^2 w^4 + 3b^3 w^3 + 3b^4 w^2 + b^5 w + b^6.$$

Comparing the inventory in the last example, focusing on the coefficients of $b^3 w^3$ we see that it is here that the number of colorings was reduced by one. Let's try three colors:

$$P(b+w+r, b^2+w^2+r^2, b^3+w^3+r^3, b^6+w^6+r^6)$$

$$= \frac{1}{12}((b+w+r)^6 + 2(b^6+w^6+r^6) + 2(b^3+w^3+r^3)^2$$

$$+4(b^2+w^2+r^2)^3 + 3(b+w+r)^2(b^2+w^2+r^2)^2)$$

$$= w^6+rw^5+bw^5+3r^2 w^4+3brw^4+3b^2 w^4+3r^3 w^3+6br^2 w^3+6b^2 rw^3$$

$$+3b^3 w^3 + 3r^4 w^2 + 6br^3 w^2 + 11b^2 r^2 w^2 + 6b^3 rw^2 + 3b^4 w^2 + r^5 w$$

$$+3br^4 w + 6b^2 r^3 w + 6b^3 r^2 w + 3b^4 rw + b^5 w + r^6 + br^5 + 3b^2 r^4$$

$$+3b^3 r^3 + 3b^4 r^2 + b^5 r + b^6.$$

Example 4.40 *This example deals with counting distinct graphs. If two vertices in the graph are connected by an edge, we will consider that edge colored black. If they are not connected by an edge, we will consider that (nonexistent) edge painted white. Let's count the number of graphs with five vertices. Before we can do this we need to count the number of edges it can have, but this is not difficult.*

$$\text{The number of edges will be } \binom{5}{2} = 10.$$

Therefore, the number of graphs possible is $2^{10} = 1024$. In general,

$$\text{the number of graphs with n vertices is } 2^{\binom{n}{2}}.$$

Now some of these graphs with five vertices are not distinct. For instance, the graph containing only the edge connecting vertices 1 and 2 is the same as the graph containing only the edge connecting vertices 3 and 4. What we really want to count is the number of distinct (or in this case we say **non-isomorphic**) graphs. What then do we mean by distinct in this case? What we mean is there is no permutation of the vertices which preserves all the edge connections, i.e. fixes the coloration. So it boils down to counting something we already know how to do.

Let's start with an easier set of graphs, namely one with four vertices. So our set X is the collection of all graphs with four vertices (there are $2^6 = 64$). Now the entire collection of permutations, S_4, is acting on the set vertices of the graphs in X. We require a table of all the possible cycle types in S_4 and the number of each type (we already saw this in Example 4.33).

Cycle Type	Number
$(*)(*)(*)(*)$	1
$(*)(*)(**)$	$\frac{4!}{2!\cdot 2} = 6$
$(*)(***)$	$\frac{4!}{3} = 8$
$(**)(**)$	$\frac{4!}{2!\cdot 2\cdot 2} = 3$
$(****)$	$\frac{4!}{4} = 6$

Since X is really the collection of colorations of edges, we need to find the corresponding permutation of the edges in order to create Polya's Formula. We shall denote the collection of edges by $\overline{12}, \overline{13}, \overline{14}, \overline{23}, \overline{24}, \overline{34}$, where $\overline{mn}$ means the edge connecting the vertices numbered m and n. Let's add one more columns to our table using a generic permutation of each vertice cycle type.

Vertice Cycle Type	Number	Edge Cycle Type
$(1)(2)(3)(4)$	1	$(\overline{12})(\overline{13})(\overline{14})(\overline{23})(\overline{24})(\overline{34})$
$(1)(2)(3\ 4)$	6	$(\overline{12})(\overline{13}\ \overline{14})(\overline{23}\ \overline{24})(\overline{34})$
$(1)(2\ 3\ 4)$	8	$(\overline{12}\ \overline{13}\ \overline{14})(\overline{23}\ \overline{24}\ \overline{34})$
$(1\ 2)(3\ 4)$	3	$(\overline{12})(\overline{13}\ \overline{24})(\overline{14}\ \overline{23})(\overline{34})$
$(1\ 2\ 3\ 4)$	6	$(\overline{12}\ \overline{23}\ \overline{34}\ \overline{14})(\overline{13}\ \overline{24})$

Therefore, Polya's Formula is

$$\frac{1}{24}(x_1^6 + 9x_1^2x_2^2 + 8x_3^2 + 6x_2x_4),$$

and the number of distinct graphs with four vertices is

$$\frac{1}{24}(2^6 + 9 \cdot 2^2 \cdot 2^2 + 8 \cdot 2^2 + 6 \cdot 2 \cdot 2) = 11.$$

We can exhibit the inventory of distinct graphs with four vertices by evaluating

$$\frac{1}{24}((b+w)^6 + 9(b+w)^2(b^2+w^2)^2 + 8(b^3+w^3)^2 + 6(b^2+w^2)(b^4+w^4)),$$

but since we only care about existing edges we can replace w by 1 and evaluate

$$\frac{1}{24}((b+1)^6 + 9(b+1)^2(b^2+1^2)^2 + 8(b^3+1^3)^2 + 6(b^2+1^2)(b^4+1^4))$$

$$= 1 + b + 2b^2 + 3b^3 + 2b^4 + b^5 + b^6.$$

In the Figure 4.19 we display the distinct graphs.

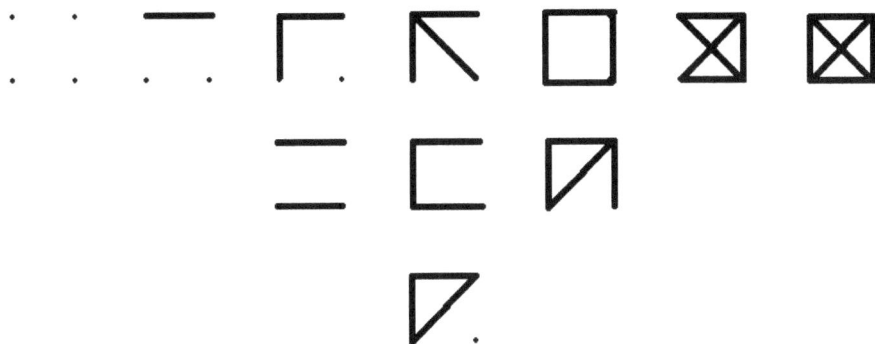

Figure 4.19 The distinct graphs with four vertices.

EXERCISES

1 List the 14 distinct colorings in Example 4.34.2

2 Compute by hand the additional fixators in Example 4.34.3.

3 Which two of the 14 colorings in Example 4.34.2 are now being equated in Example 4.34.3?

4 Investigate the case of graphs with five vertices as we did with four in Example 4.40.

5 Consider again the problem where we wish to paint the roof of a house (see Figure 4.20). There are four sections of roof each of which can be painted in one of two colors: sienna and ochre. Our goal is to investigate the number of distinct colorings up to (two) rotations and (two) reflections.

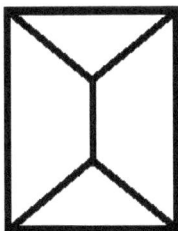

Figure 4.20 The roof of a house.

a. Create the Polya Polynomial associated with this problem.

b. Use part b. to count the number of distinct colorings.

c. Use part b. to list the inventory of colorings.

d. Exhibit an example of each of the distinct colorings.

References

[1] G. Cantor. Uber eine elementare frage der mannigfaltigkeitslehre. *Jahresbericht der Deutschen Mathematiker-Vereinigung*, 1:75–78, 1891.

[2] H. D. Ebbinghaus, J. Flum, and W. Thomas. *Mathematical Logic*. Academic Press, NY, USA, 1st edition, 1977.

[3] H. B. Enderton. *Elements of Set Theory*. Springer-Verlag, NY, USA, 1st edition, 1984.

[4] S. Galovich. *Doing Mathematics*. Saunders College Publishing, FL, USA, 1st edition, 1993.

[5] G. H. Hardy. *A Mathematician's Apology*. Cambridge University Press, NY, USA, 1st edition, 1987.

[6] A. N. Kolgomorov and S. V. Fomin. *Introductory Real Analysis*. Dover Publications, Inc., NY, USA, revised english edition, 1975.

[7] W. LeVeque. *Fundamentals of Number Theory*. Addison-Wesley Publishing Company, Inc., 1st edition, 1977.

[8] J. Murzi and B. Topey. Categoricity by convention. *Philos Stud*, 178:3391–3420, 2021.

[9] T. Sundstrom. *Mathematical Reasoning: Writing and Proof*. CreateSpace Independent Publishing Platform, 2.1 edition, 2014.

Index

alternating group, 129
antecedent, 3

Cantor's Continuum Hypothesis, 121
Cartesian product, 72, 90
computability, 59
conclusion, 3
consequent, 3
countable, 110
counterexample, 13

DeMorgan's Laws, 8
denumerable, 110
dihedral group, 133
disjoint union, 26
Division algorithm, 41, 59, 68, 103

element chasing, 24, 27
equivalence class, 84
    quotient set, 84
    representative, 84
Euclidean algorithm, 62
Extended Euclidean algorithm, 64

function, 90
    bijection, 92
    branch, 49, 93
    codomain, 90
    composition, 96
    domain, 90
    equality, 90
    extension map, 94
    identity map, 94
    image, 90

inclusion map, 94
injective, 92
input, 90
inverse, 97
inverse image, 99
one-to-one coorespondence, 92
output, 90
permutation, 94
piecewise defined, 49, 93
projection map, 94
quotient map, 94
range, 90
real-valued, 91
restriction map, 94
vertical line test, 91
well-defined, 99
Fundamental Theorem of Arithmetic, 43, 69, 105

greater than, $>$, 18
greatest common divisor, 60, 104
group
    action
        transitive, 141
    symmetric group, 94

Hasse diagram, 82
hypothesis, 3

Inclusion-Exclusion Principle, 76
induction
    atomic step, 37
    base case, 37
    inductive step, 37
    strong induction, 43

weak induction, 37
integers
    congruence, 80
    modulus, 80

Law of the excluded middle, 32, 56
logical connective, 2
    and, 2
    equivalence, 2
    exclusive or, 2
    implication, 2
    not, 2
    or, 2

map, 91
membership function, 13
modus ponens, 6

number systems
    integers, 13
      divides, 19
      even, 18
      odd, 18
    irrational numbers, 13
    natural numbers, 13
      factorial, 61
      prime, 34
      Well-ordering principle, 35
    rational numbers, 13
    real numbers, 13
      absolute value, 49
      triangle inequality, 55

ordered pair, 72

partition, 86
Polya's Formula, 146
premise, 3
prime, 104
proof
    contrapositive, 30

    direct, 17
    indirect, 31
proposition, 1

quantifier
    existential, 12
    universal, 12

real numbers
    interval, 73
relation, 79, 90
    anti-symmetric, 80
    congruence, 106
    equivalence relation, 81
    function, 81
    irreflexive, 80
    partial ordering, 81
    reflexive, 80
    symmetric, 80
    transitive, 80
    trichotomy, 81

sentence, 12
set, 21
    cardinality, 23, 110
    complement, 26
    difference, 26
    disjoint, 30
    dominance, 115
    element, 21
    empty set, 23
    equality, 23
    equinumerous, 115
    finite, 21
    infinite, 21
    intersection, 26
    null set, 23
    partially ordered, 81
    power set, 40
    proper subset, 23
    roster, 21

set builder, 21
size, 21, 23
strict dominance, 116
subset, 23
symmetric difference, 26
union, 25
universe, 26
statement, 1
and, 2
conditional, 3
contradiction, 6
contrapositive, 8, 10
converse, 10
direct, 10
equivalence, 4
exclusive or, 3
implication, 3
inverse, 10
logical equivalence, 6
necessary, 11
or, 3
sufficient, 11
tautology, 6

symmetric group on n,
94
k-cycle, 124
action, 136
fixator, 137
orbit, 137
stabilizer, 137
conjugacy class, 138
coset, 138
cycle type, 127
disjoint cycles, 125
order, 130
transposition, 124
even and odd,
128

truth table, 2

uncountable, 110
unit, 107

variable, 12

zero divisor, 107

For Product Safety Concerns and Information please contact our EU
representative GPSR@taylorandfrancis.com
Taylor & Francis Verlag GmbH, Kaufingerstraße 24, 80331 München, Germany

www.ingramcontent.com/pod-product-compliance
Lightning Source LLC
Chambersburg PA
CBHW070727220326
41598CB00024BA/3331

9 781032 687704